TEACHING OF ENVIRONMENT

New Trends and Innovations

TEACHING OF ENVIRONMENT

New Trends and Innovations

Edited by
P.P. SINGH
SANDHIR SHARMA

DEEP & DEEP PUBLICATIONS PVT. LTD.
F-159, Rajouri Garden, New Delhi - 110 027

TEACHING OF ENVIRONMENT

ISBN 81-7629-615-5

Printed in India at NEW ELEGANT PRINTERS
A-49/1, Mayapuri, Phase-I, New Delhi - 110 064

Published by DEEP & DEEP PUBLICATIONS PVT. LTD.,
F-159, Rajouri Garden, New Delhi - 110 027 • Ph. : 25435369, 25440916
E-mail : deep98@del3.vsnl.net.in • ddpbooks@yahoo.co.in
Showroom :
2/13, Ansari Road, Daryaganj, New Delhi - 110 002 • Telefax : 23245122

CONTENTS

Preface

The technological innovations that produced, first, the Agricultural Revolution, then, the Industrial Revolution, and, in the latter part of the twentieth century, the Information Revolution have enabled people not only to manipulate the natural environment, but to do so on an ever grander scale and at an ever accelerating pace. The use of biotechnology and other new technologies undoubtedly will enable people to make even more extensive and more rapid changes in their environment in the future.

As long as there existed on earth uninhabited or sparsely inhabited lands and as long as the impact of human technological innovation on the environment was confined to small segments of the planet, the environmental consequences of the human use of technology and human population growth could be largely ignored. Today, though, there are few frontiers left on earth, and many human activities have global impacts. Chemicals released into the atmosphere cause acid rain that destroys forests hundreds of miles away from the point of source and may even affect the climate of the whole earth; the fishing and ocean dumping practices of one nation impinge on the availability of seafood to nations around the globe; microbes on food produced in one country result in outbreaks of illness thousands of miles away; people caught in the crossfire of war or facing environmental degradation in their homelands may emigrate extremely rapidly en masse to very distant parts of the world. Hence, all people are being forced to confront the limitations of their environment – the earth.

The rise of concern and consciousness about environmental issues in the developing countries can be understood only in the broader context of international movement in general and the domestic impulses that drive policy formulations related to economic development in particular. These countries are faced with meeting the twin goals of economic and environmental development. They do not have the luxury to choose one over the other. The policy choices to accomplish both are not easy. Environmental decay resulting from population explosion, urban degradation, tropical deforestation, ozone depletion, and hazardous waste disposal is real, but so is the need for sustaining basic needs of day-to-day life. Their dilemma and frustration is often times expressed in their policy stands taken at various international forums.

This book would be of use to all interested in teaching of environmental studies.

We are highly thankful to Dr. K.N.S. Kang, Director, Punjab College of Technical Education for providing us the motivation to work on this project. We are grateful to all others who have contributed for the preparation of this book. We acknowledge with thanks the help of various librarians and authors of various books from whom we have heavily drawn upon the material of several publications. We would also like to thank the publishers for taking up this book for publication and bringing it out in time.

P.P. SINGH
SANDHIR SHARMA

1

TEACHING OF ENVIRONMENT SCIENCE

Introduction

The environment looms large on the global political agenda. The increased attention is the result of several factors: increased awareness of the environmental damage caused by economic activity; rapid population growth; the perception of the earth as a single biosphere; and, more recently, the end of the Cold War. During the past few decades, the scientific community has uncovered new information on the environmental consequences of humankind's interaction with the earth. Much of this has found its way into the popular media. As a consequence, the public has some familiarity with issues such as global warming, ozone depletion, nuclear pollution, and hazardous waste disposal. This period has also seen accelerating population growth, growth that has thrown into stark relief the problems caused by the inadequate supply of food and resources in certain localities and by the maldistribution of food on a global level.

In the past, food and resource shortages were perceived primarily as local or national problems and, consequently, as the responsibilities of many different jurisdictions. But this has changed because of the new perception of the earth as a single biosphere. There is an awareness that actions in one part of the world can have implications for resource availability in distant areas. For example, a poorly managed nuclear power plant in one country can damage people and crops thousands of miles away. Similarly, each country's decision regarding the use of fossil fuels will have implications for global warming and for life on earth. As a consequence, many issues that were initially regarded as matters of local concern are increasingly becoming issues of global concern and, therefore, the subject of international politics. Finally, the end of the Cold War has allowed the environment to rise on the global agenda. As military security issues loom less large on that agenda, more attention is paid to issues of environmental security. This change was illustrated by the unprecedented gathering of 118 heads of state at the United Nations Conference on Environment and Development in Rio de Janeiro in June 1992.

As the scientific community has uncovered more about the causes of commons environmental problems such as global warming and ozone depletion, the new findings have highlighted the unified nature of the biosphere. Since an action by one state can have environmental implications for the rest of the global community, it is no longer acceptable for countries to defend their actions with claims of sovereignty. Gaining ascendancy is the perspective that sovereignty as a legal right of states will need to be modified in the light of these new circumstances. Increasingly, it has become obvious that the existing institutional arrangements are inadequate to handle the burgeoning environmental problems, and the global community of states has begun the process of establishing new institutions for global environmental management. For the most part, this process has gone forward in a piecemeal manner, with separate regimes for specific issue areas. Some areas have regimes that include specific control measures, and other areas are yet to be carefully addressed. The formation and evolution of environmental regimes have cast into stark relief the differences between the interests of the Third World and those of the industrialized states, a difference that is linked to the two groups' positions in the world economy.

The Economic Context

It is essential to examine the economic context within which environmental practices and policies evolve. Economic growth is the lodestar for much of the world. Regardless of their economic status and political perspective, a primary objective of nation-states is growth. However, current economic development strategies can have significant consequences for the environment. One important measure of development is per capita energy usage. In the past few decades, the reliability of this indicator has been affected by increased energy efficiency in some states, but economic growth is still an energy-intensive process. Between 1965 and 1990, the commercial energy consumption per capita increased from 936 to 1,316 kilograms of oil- equivalent energy. This does not include energy usage in the noncommercial sector, which is significant in some developing countries.

Two major ecological problems associated with energy consumption are resource depletion and pollution. Fossil-fuel sources of energy, such as oil and coal, are finite; they also pollute the air and water and contribute to global warming. A third fossil-fuel source, wood, also contributes to global warming and atmospheric pollution. Although in theory wood is a renewable source, in some parts of the world wood harvesting has passed the maximum sustainable level. Nuclear power is sometimes touted as the solution to the problems of energy scarcity and global warming, but the use of this form of energy raises concerns regarding public safety and waste disposal. Even a renewable source like hydroelectric power is not without environmental problems. Mega-dam projects sometimes displace thousands of people and destroy ecosystems.

The increasing international trade, which is such a vital component of our global economy, also has important environmental implications. For the past 40 to 50 years, great emphasis has been placed on the desirability of free trade. However, trade is a resource-consumptive process; the more economies become dependent on far-flung trade, the more resource consumptive trade will become. Regional trading blocs seem to be a good idea, but they are being used to supplement, rather than supplant, global trade. The various regimes that manage global and regional trade also impose specific environmental costs. One example of this is the pressure to standardize health and environmental guidelines, often to a level below those already established in many developed and developing countries.

The environmental consequences are inevitable, given the logic of a world economy that relies on infinite growth. At the 1993 Group of Seven Summit, one of the few things the leaders were able to agree on was the importance of economic growth. But in our global economic system, growth results in the use of finite resources, in the pollution caused by the use of these resources, and in the problems associated with the disposal of the hazardous wastes that are often the by-products of industrial growth.

The benefits of economic growth, such as they are, come at great cost in terms of resource depletion and pollution. But these costs and benefits are not shared equally among the world's peoples. Although Third World countries have limited access to these benefits, they pay a disproportionate share of the costs. In large part, developing countries' role in the global economy is rooted in their colonial past; dependency theorists see the development and wealth of the North and the underdevelopment of the South as functions of each other. Patterns of resource exploitation emerged that enabled the colonial power to accumulate capital and assume political hegemony over the subordinate colony, leading to a pattern in which the colony's resources were used in a manner incongruent with local needs. At the same time, these patterns of resource exploitation incorporated new products and technologies from other areas of the globe. Developing countries became very dependent on both export and import trade and decolonization did not end this pattern.

As a result of the internationalization of economies, developing countries often have only minimal control over the disposition of their resources. In addition, the degree and pattern of use of these resources do not directly reflect the needs of the source country. Outsider interests such as transnational corporations and consumers in other countries often have more influence over resource choices than do the citizens of the particular nation state. Although foreign and international interests may also be important in developed countries, the size and precarious economic health of many of these countries' economies make them more vulnerable to outside interests.

Because of the Third World's position in the global economy, on a per capita basis these countries are generally less resource consumptive and less polluting of the atmosphere than the industrialized countries. But the Third World countries hope their economic position is temporary. They expect further industrial growth, and, given the available industrial processes, this is likely to have negative ecological implications. In addition, the poverty in many of these countries has created pressures on natural resources, sometimes resulting in the despoliation of natural ecosystems such as forests and coastal areas, with significant implications for biological diversity.

The political and economic context has affected the way in which the developing countries participate in environmental politics, with the end of the Cold War further modifying their participation. During the Cold War, some of these countries used to play First World against Second World interests and were sometimes able to benefit economically from this maneuver; with the end of the Cold War they have lost this leverage, and the interests of the industrialized world have shifted to the development of the resources and markets of the former Soviet bloc countries. The end of the Cold War has put global environmental politics higher on the global agenda, but in this post—Cold War context Third World countries have lost some leverage, calling into question their ability to affect the evolution of global environmental politics.

Global Environmental Politics in the Twentieth Century

Concern about environmental degradation is not a twentieth-century phenomenon. In the second century B. C. , Plato talked about the erosion caused by earlier deforestation in Attica, and in the third century B. C., Erastothenes described how governmental land policy, navigation needs, and mining resulted in the deforestation of Cyprus. But for the most part, the concerns expressed throughout the centuries did not generate any widespread alarm. The first vestiges of an environmental movement did not develop until the second half of the nineteenth century. In 1865 the first private group dedicated to environmental protection, the Commons, Footpaths, and Open Spaces Preservation Society, was established in Britain. A few decades later, groups focusing on wilderness preservation and resource conservation developed in the United States.

Environmental Politics, 1900-1960

Rudimentary efforts at international organization took place early in the twentieth century. In Paris in 1909, at a meeting of the International Congress for the Protection of Nature, European nature protectionists proposed the creation of an international nature protection body. When this idea was proposed to the governments of 19 major states, all but two agreed in principle. A Consultative Commission for the International Protection of Nature was established in 1913; its stated purpose was the collection, classification, and publication of information on the international protection of nature. However, this early institutional attempt was killed by the outbreak of World War I.

Many of the early efforts at international cooperation were regional in scope and dealt with issues such as migratory animals and wildlife. They included such agreements as the European Convention Concerning the Conservation of Birds Useful to Agriculture, which was signed by 11 nations in 1902; and the Convention on Nature Protection and Wildlife Preservation in the Western Hemisphere in 1940, which was sponsored by the Pan American Union.

Much of the postwar development of international environmental politics has taken place under the auspices of the United Nations (UN). The United Nations Educational, Scientific, and Cultural Organization (UNESCO), which was established in 1945, and the International Union for the Protection of Nature (IUPN), which was formed in 1948, were important actors in this political process. The postwar activity led to an international conference on the conservation and use of resources, called by the United Nations Economic and Social Council (ECOSOC) for summer 1949.

This first major international conference on the environment was held in the United States. The UN Scientific Conference on the Conservation and Utilization of Resources (UNSCCUR) was organized by the Food and Agriculture Organization (FAO), the World Health Organization (WHO), UNESCO, and the International Labour Organisation and was attended by more than 500 delegates from 49 countries. The delegates represented themselves and were not official representatives of their countries. In the Cold War climate, there were no delegates from the Soviet bloc; in fact, more than half of the delegates were from the United States. It was a meeting of scientists rather than politicians or diplomats, and it was criticized for taking place within a political vacuum. UNSCCUR's mission was to address the problem of nature protection primarily from an economic perspective. The conference had no power to bind governments or make recommendations to them. Although questions of policy did surface, the conference's primary task was to share information on techniques for the conservation and utilization of resources. The scientists were generally confident that the world could support its rapidly growing population. The prevailing opinion was that technology could successfully address any obstacles. The meeting focused on the environment as a resource base; therefore, it addressed global resource rather than pollution issues. The resource areas examined included minerals, fuels and energy, water, forests, land, and wildlife and fish. The conferees also looked at the integrated development of river basins. Although this conference focused on science rather than policy, some policy analysts regard it as the first major event in the development of the international environmental movement.

Work on environmental protection was also proceeding outside of the United Nations within the framework of the IUPN, which was a hybrid of governmental and nongovernmental bodies. Its focus was the preservation of wildlife and the natural environment, education, scientific research, legislation, and the collection, analysis, and dissemination of data and information. Concurrent with UNSCCUR, IUPN held the International Technical Conference on the Protection of Nature (ITC), which was sponsored jointly with UNESCO. Although the conference focused on the protection of nature, it also addressed the conservation of renewable natural resources. Unlike UNSCCUR, ITC had the task of making recommendations concerning national and international legislative measures. The conference agenda also included the matter of technical assistance to less developed countries. It was attended by representatives from 32 countries and 11 international organizations, including FAO, WHO, the International Council for Bird Preservation, the International Council of Scientific Unions, and the Organization of American States. At this time there were fewer than 40 independent developing country-states. Most of today's Third World states were still colonies, so many of the conservation issues that related to these areas were addressed by the industrialized countries. Preparatory meetings and the conference itself were dominated by the developed countries; the independent developing states had very little impact on the process.

Between 1948 and 1956, IUPN's agenda gradually broadened to include conservation. This was reflected in its new name—the International Union for Conservation of Nature and Natural Resources (IUCN). Its agenda now included the conservation of representative ecosystems.

Environmental Politics, 1960 to Stockholm (1972)

The early 1960s marked the period of greatest change in both the environmental movement and environmental politics. Books such as Rachel Carson 's *Silent Spring* and Rolf Edberg's *On the Shred of a Cloud* mobilized public opinion. The environment moved from being the concern of a few scientists, administrators, and conservation groups to being the focus of a mass movement that affected much of the industrialized world. Consequently, the public began to pressure governments to address issues of pollution and resource mismanagement. Sharpening the focus on the environment was the growing concern about the potential impact of nuclear fallout.

Public concern was heightened by a series of environmental disasters between 1966 and 1972. Among these was the wreck of the *Torrey Canyon* in March 1967. Hundreds of miles of Cornish coastline were polluted when about 875,000 barrels of crude oil were spilled. Two years later, an oil platform blowout off the coast of Santa Barbara, California, polluted miles of California coastline. Japan was the site of another well-publicized disaster, which had more direct human implications. Local residents were poisoned as a result of decades of mercury discharges into the ocean. Fifteen hundred people became sick, and more than 200 died. Although environmental disasters were not new phenomena, the increased public sensitivity to environmental issues after the early 1960s heightened the impact of these disasters.

In addition, this period also saw an increase in scientific knowledge. It is within this context that the Biosphere Conference

(the Intergovernmental Conference of Experts on the Scientific Basis for Rational Use and Conservation of the Resources of the Biosphere) of 1968 was held in Paris. The conference was primarily a UN-sponsored effort; it was organized by UNESCO, with help from FAO and WHO. The International Union for the Conservation of Nature and Natural Resources and the International Biological Programme were also involved in its organization. The Biosphere Conference addressed the human impact on the biosphere, including the effects of air and water pollution, overgrazing, deforestation, and the drainage of wetlands. Conference participants emphasized that environmental deterioration had reached a critical threshold and that methods of developing and using natural resources had to change. This was clearly a change from UNSCCUR's general optimism 19 years before. The conference recommended additional research on ecosystems, human ecology, pollution, and genetic and natural resources; new approaches to environmental education; and assessments of the ecological impacts of large-scale development projects.

Some of the themes explored at the Biosphere Conference were developed further in the United Nations Conference on the Human Environment (the Stockholm Conference) in 1972. However, whereas the Biosphere Conference focused on the scientific aspects of environmental problems, the Stockholm Conference looked at the broader political, social, and economic issues. As a result, nongovernmental organizations (NGOs) were involved at Stockholm to a greater extent than had been the case at the Biosphere Conference; in fact, observers from 258 NGOs were present. Participants at Stockholm had a very broad agenda, reflecting the wide-ranging priorities among the members of the international community.

The Stockholm Conference was the result of efforts to place the protection of the biosphere on the official agenda of international policy. The participation of most of the world's nations and many organizations suggests that these efforts succeeded. There were delegates from 113 nations, 21 UN agencies, and 16 intergovernmental organizations. The conference took place at a time when issues of global inequality were gaining ascendancy on the international agenda; it is not surprising, then, that North-South differences over global economic relations and environmental politics were major elements of this conference. Third World countries played a prominent role in pre-conference discussions and thus were able to modify the agenda.

The conference had been planned to address the particular set of environmental problems identified by developed countries, but the developing countries were concerned that the conference outcome might have negative consequences for their development prospects. The industrialized countries were persuaded to broaden their environmental agenda to include issues of particular interest to the Third World such as shelter, food, and clean drinking water. The tendency had been to focus on issues such as pollution, population explosion, conservation of resources, and limits to growth. Some Third World countries saw much of this discourse as an academic exercise, with little relevance to their everyday lives. Using their United Nations General Assembly voting power, they were able to press developed countries to embrace a broader vision of environmental issues, one that addressed Third World development needs. Third World countries were beginning to play a role in determining the international environmental agenda.

The most tangible outcome of the Stockholm Conference was the creation of the United Nations Environment Programme (UNEP). This institutional framework would address the issues put forward in the Stockholm Declaration, Principles and Action Plan. In the decades since Stockholm, it has played a vital policy-coordinating role. The developing countries campaigned to have the UNEP secretariat located in a developing country. Their position prevailed, and the secretariat was placed in Nairobi, Kenya.

Environmental Politics, Stockholm to Rio (1972-1992)

The second wave of international environmentalism was spurred by events such as the Bhopal disaster at the Union Carbide plant in India, the Chernobyl explosion, and the Exxon *Valdez* oil spill. On the research front, there was growing consensus among scientists regarding environmental issues such as ozone depletion and climate change. In addition, renewed attention was given to the tension between growth and the environment, one example of which was the 1987 report by the World Commission on Environment and Development. The report concluded that continued economic growth was essential, particularly for the people of the developing world, but it made the point that this growth would not be feasible if traditional growth patterns were pursued and called for a switch to new modes of environmentally sound or sustainable development. It was within the context of this second wave of environmentalism that the UN General Assembly mandated the 1992 UN Conference on Environment and Development (the Earth Summit), which was held in Rio de Janeiro, Brazil.

The Earth Summit was held on the twentieth anniversary of the Stockholm Conference. It was initiated by developed Western nations to deal primarily with the environmental consequences of industrialized society. It followed nearly two years of preparatory work. More than 170 nations sent delegates to the conference; also present were representatives of thousands of NGOs, many major industrial corporations, and nearly 9,000 members of the media. The conference drew 118 heads of state, the largest such gathering ever.

Yet the conference proved that the world was still divided. For Rio, the Third World and the industrialized countries had suggested agendas that reflected their differences at the Stockholm Conference. The developed countries wanted to address ozone depletion, global warming, acid rain, and deforestation, whereas the developing countries were more interested in examining the relationship between developed countries' economic policies and developing countries' sluggish economic growth. Leaders of developing countries insisted that environmental protection alone was not enough; therefore, any agreement about global environmental issues must also include measures for economic development. The sentiment was that an environmentally healthy planet was impossible in a world that contained significant inequities. The North acknowledged the connection between environment and development in many of the conference documents, and some industrialized countries did promise increased monetary aid, but the North refused to commit to specific steps to reduce its own industrial pollution and continued to press the developing countries to protect their forests and wildlife.

The Earth Summit produced new international treaties on climate and biodiversity, a statement on forests, the Rio Declaration, and an action plan for sustainable development, known as Agenda 21. In addition, the meeting produced a consensus that preparations should begin for a treaty to curb desertification. The parties also established a watchdog commission to monitor progress on the pledges made at Rio.

The Purpose of the Book

As nation-states of the global community negotiate global regimes, it becomes clear that their common concern for the biosphere does not necessarily lead to shared interests. Developing countries recognize different priorities from those identified by the developed countries. This book examines the role of the Third World in global environmental politics. Specifically, it examines how these countries have attempted to modify the evolving global environmental regime and assesses the extent to which they have been successful. It addresses these issues within the context of globalization, both economic and ecological. It finds that whereas the former acts as a constraint on developing countries' influence in environmental politics, in certain contexts ecological globalization provides them with some room for maneuver. The focus on globalization of the economy pays particular attention to the constraints imposed on developing countries as a consequence of their role in the world economy. Since the ready assumption is that the industrialized countries hold the preponderance of power, it is easy to assume that the developing countries have little or no influence on the formation of environmental regimes. This would be consistent with the realist perspective of international politics.

For realists, international behavior is based primarily on interests and power. Politics is a struggle for power within a state-centric system. Consequently, the realist paradigm is based on conflict and competition among states. It assumes that states are unitary actors with gradations of capabilities among them and perceives a clear separation between domestic and foreign policy. The regime approach, which is central to this study, makes several modifications to the traditional realist perspective. It assumes that competing, self-interested states can find it in their interest to opt for cooperative behavior. In making this assumption, it modifies the realist perspective regarding the role played by self-interest. The regime concept also embraces a more expansive conception of power. As with realist theory, the regime approach sees states as primary actors, but unlike realist theory, it also recognizes that regimes can be major global actors.

Global and regional environmental problems present challenges that cannot be clearly understood using the state-based theory of realism. Environmental problems often transcend national boundaries, and the institutions and organizations that are emerging as critical for addressing these issues include regimes and international organizations, both of which are neglected in realist theory. Additionally, realist theory does not allow for the focus on cooperation called for by an examination of regimes. Scholars find the regime approach particularly useful because of its focus on the cooperative behavior that is an important part of the evolution of global environmental politics. The study of regimes offers a different perspective on power. The very nature of global environmental problems creates new power dynamics between developed and developing countries and changes the way in which gradations of power among countries are perceived. For example, the fact that the Third World is likely to release immense amounts of greenhouse gases in its efforts to industrialize provides these countries with an important bargaining chip. The globalization of the environment implies that biosphere issues are considered within a commons context, in which situation any actor has the ability to subtract from the welfare of the other actors. This, then, is a potential basis for leverage by the Third World.

Although the globalization of the environment and of the economy both result in interdependence, they represent two different types of interdependence. Economic interdependence is a constructed interdependence since it is a result of the current economic and social system. In a world in which there is globalization of resources, capital, and labor, there is an increasing number of transnational, intergovernmental, and trans-governmental relationships. The resulting interdependence makes all global actors, to a greater or lesser degree, more sensitive to changes in other actors and more vulnerable to external events. Although the multiple linkages are the most pronounced among the industrialized countries, major intergovernmental organizations

(IGOs) and NGOs have been instrumental in integrating the Third World countries into the global economy. Because of the power asymmetry, the sensitivity and vulnerability that are consequences of an interdependent world are more likely to be reflected in the Third World as costs rather than benefits. Whereas this interdependence is likely to be a long term feature, it is not a permanent, unchanging characteristic of the world system. Ecological interdependence, however, is a consequence of the biosphere within which all human beings live. It is a naturally occurring interdependence, and all of humankind must operate within the constraints of the biosphere.

The concept of a biosphere emerged in the late nineteenth century; however, much of the credit for the development of this concept has been given to Russian mineralogist V. I. Vernadsky, who wrote about it in the first half of the twentieth century. The biosphere is composed of the lower atmosphere, the oceans, and the lithosphere. Vernadsky argued that outside of this "terrestrial envelope" humankind could not live naturally. He also asserted that because of the natural and necessary connections among all living organisms, no organism is truly free of other organisms; there is a natural interdependence.

In a system in which there is economic interdependence, economic asymmetry can be exploited to benefit the more powerful actors. However, the regimes examined in this book illustrate that a general perception of ecological interdependence places constraints on both the powerful and the puny. In situations in which developed country actors perceive this interdependence, Third World actors are able to exercise some modest leverage in environmental politics. This book explores how this power is exercised and how it is modified, and it focuses on the role of the Third World in the formation and evolution of three environmental regimes: the hazardous waste trade regime, the biodiversity regime, and the ozone protection regime.

Each of these regimes has been developed to manage a different kind of environmental problem. Although both the biodiversity and ozone protection regimes address common property resource issues, there are substantial differences between these two commons problems. Since ozone protection is an atmospheric issue, we are dealing with a resource held in an open access arrangement; biodiversity, on the other hand, is held in a mixed property regime with a wide variety of arrangements, including private property, government control, communal property, and open access. The hazardous waste trade regime addresses a trans-boundary problem rather than a commons problem. Its purpose is to regulate the shipment of wastes from one country to another. These varying property and access arrangements affect how states perceive their interdependence and vulnerability; consequently, they have implications for the leverage of the Third World in international environmental negotiations. This study focuses on the role of the Third World at three stages of regime evolution: problem definition, bargaining, and regime transformation.

2

BIODIVERSITY

Biodiversity or Biological diversity pertains to the variety of species including plants, animals, human beings, microorganisms and ecosystems in a region. country, continent or the entire globe. Biodiversity comprises every form of life from the tiniest microbe to the mightiest beast or tree.

Existence of Species

The greatest challenge to mankind is perhaps the estimation of biodiversity, for it would require millions of man days of painstaking work by highly trained biologists.

The number of living organisms on earth has been roughly estimated to be between 5 and 50 million or more, though only about 1.4 million have been described, of which 750,000 are insects, 40,000 are invertebrates, 250,000 are plants and 360,000 pertain to the microbiota. According to some recent estimates, the number of insects alone may be as high as 10 million, but many believe it is more likely to be around 5 million. Most of the undescribed species are in the rain forests.

There is a tremendous difference in the number of species belonging to each category of organism.

Relative number of described species in major taxa:

Taxon	*No.of described species*
Monera (bacteria, blue-green algae)	4,760
Fungi	46,983
Algae	26,900
Plantae (multicellular plants)	248,428
Protozoa	30,800
Porifera (sponges)	5,000
Coelenterata (jellyfish, corals, comb jellies)	9,000
Platyhelminthes (flatworms)	12,200
Nematoda (roundworms)	12,000
Annelida (earthworms, etc.)	12,000
Mollusca (mollusks)	50,000
Echinodermata (starfish etc.)	6,100
Insecta	751,000
Non-insect Arthropoda (mites, spiders, crustaceans, etc.)	123,161
Pisces (fish)	19,056
Amphibia (amphibians)	4,184
Reptilia (reptiles)	6,300
Aves (birds)	9,040
Mammalia (mammals)	4,000

Types of Biodiversity

Biodiversity can be studied at three different levels : genetic, species and ecosystem.

Genetic Diversity

It is concerned with the variation in genes within a particular species. Many billion individuals are produced through sexual reproduction. Each differs from the other in its hereditary constitution or the genetic information contained in its genes. It is this difference which has given us beautiful butterflies, roses, parakeets or coral in a myriad hues, shapes and sizes.

Examples: Brown Bear, Sloth Bear, Himalayan Black Bear; Himalayan Tahr, Nilgiri Tahr

Species Diversity

It refers to the variety of living organisms on earth. Species differ from one another markedly in their genetic makeup and do not interbreed in nature. Closely related species however have much of their hereditary material in common. For instance, humans and chimpanzees have about 98.4 per cent of the same genes.

Ecosystem Diversity

This refers to the variety of habitats. A habitat is the sum total of the climate, vegetation and geography of a region. There are several kinds of habitats around the world. Change in climatic conditions is accompanied by a change in vegetation as well. For instance cactus and thorny scrub replace grassland or forests as the climate changes from temperate to hot and dry. Each species is adapted to a particular kind of environment. As the environment changes, species best adapted to that environment become predominant. Thus the variety or diversity of species in the ecosystem is influenced by the ecosystem itself. Geographical changes within an ecosystem lead to genetic diversity or diversity within the same species. In India the Himalayan Tahr is found in the Eastern Himalayas, while the Nilgiri Tahr inhabits the slopes of the southern Western Ghats.

Examples: Corals, Grasslands, Wetland, Desert, Mangrove, Tropical rain forests, etc.

The Megadiverse Nations

Biodiversity is not evenly distributed among the world's countries. Barely a dozen countries lying partly or entirely in the tropics account for 60 to 70 percent of the world's biodiversity. They are referred to as Megadiverse Nations. Species richness or diversity is high in the tropics and the cold depths of the oceans as both these regions are somewhat protected from disturbances due to inaccessibility. Rapid change in environment as at the seashore and long tracts of mountains also account for high species turnover.

The 12 megadiverse nations of the world. The countries on the megadiversity list include Brazil, Colombia, Ecuador, Peru, Mexico, Zaire, Madagascar, Australia, China, India, Indonesia and Malaysia.

Countires	*No. of mammals*	*No. of birds*	*No. of reptiles*	*No. of amphibians*	*No. of angiosperms*
Australia	255	-	686	197	23,000
Brazil	428	622	467	516	55,000
China	394	1195	270	265	27,000
Columbia	359	1721	383	407	45,000
Ecuador	280	1447	345	358	15,000
India	350	1200	353	197	15,000
Indonesia	515	1519	600	270	20,000
Madagascar	-	250	269	144	10,000
Malaysia	293	1200	294	171	15,000
Mexico	449	1010	717	282	25,000
Peru	361	1703	297	251	20,000
Zaire	409	1086	280	216	10,000

India is one of the richest nations in terms of biological diversity. India owes this to its position in the tropical and subtropical latitudes with their inherent wealth of life. Rapid change in environment occurs at the long coastline and at the edges of the deserts. India also possesses islands like the Andaman and Nicobar and Lakshadweep with their own endemic species, and mountain ranges like the Himalayas and the Western and Eastern Ghats. All these factors help to make it a megadiverse nation. Group No. of species in India No.of species in the world:

Mammals	372	4,231
Birds	1224	8400
Reptiles	408	5375
Amphibians	197	2000
Fishes	2546	23,400
Flowering Plants	15000	800,000

Two of India's great mountain ranges, the Eastern Himalayas and the Western Ghats have been designated among the world's eighteen 'hotspots' of biodiversity. The third most significant area of biodiversity in India is the island ecosystem of the Andamans and Nicobar.

No. of endemic species found in the hotspots of megadiverse India

Region	*Mammals*	*Plants*	*Reptiles*	*Amphibians*
E. Himalayas	N.A.	3500	20	25
Western Ghats	7	1600	91	84
Andaman and Nicobar	33	144	23	2

Significance of Biodiversity

Does the earth really need to have about 50 million different species? If some species are lost or become extinct, the total number would be affected only minutely. So why is the conservation of biodiversity of utmost importance today? The answer to these questions lies in the web of life that links each and every form of life on earth.

Food Chain

The building blocks of plants, animals and humans are identical, and are made up of the four elements, carbon, oxygen, nitrogen and hydrogen. These elements are available in the environment in the air, water and soil. However, only green plants can absorb nitrogen from the soil through their roots, and use sunlight and water by a process called photosynthesis to produce energy. They are known as producers. Animals and humans get their food only by consuming plants or other animals. Therefore they are known as consumers.

The chain that links consumers to producers is called the food chain or web of life. Every living creature is to be found on a food chain. There are several food chains and depending on the environment it can be complex or simple.

All life on earth is interdependent and man is only a strand in this delicate web of relationships.

Food Chains in Water

Consider a very short and simple food chain in water . Green algae are eaten by water fleas, which are in turn eaten by hydras. An example of a slightly longer chain is, plankton - krill - small fish - predator fish - seal - man.

Food Chains on Land

Insects like the caterpillars feed on green leaves which are eaten by small birds like the sparrow. Carnivorous birds like the hawk hunt the sparrows. When the hawk dies its remains are returned by other insects like the sexton beetle to the soil, where they become nutrients for the tree.

Green grass is eaten by grazing animals like the deer, bison, zebra and cattle. These are in turn eaten by carnivores like the lion, tiger, leopard and man.

Grasshoppers eat grass and are themselves eaten by frogs. Snakes eat frogs and rodents.

Fig trees are pollinated by tiny insects called fig wasps. The female wasp enters a fig flower to lay her eggs and dies. In the process she brings in fig pollen which fertilises the flower to produce the fig fruit. The eggs develop in the flower to larvae which feed on the fruit and become adult wasps that mate within the flower. The female wasp - whose life span is only two days - then leaves covered with the pollen in search of a fig in flower. Thus both fig trees in fruit and in flower should be present throughout the year. Fruit eating mammals hence have figs throughout the year and help in the dispersal of its seed just as they disperse seeds of other fruits. Therefore other plants are also dependent on the fig wasp which initiates the pollination.

All food chains are fragile. If a link in the chain is broken, it sets off a series of reactions.

If snakes are destroyed the population of rats and moles would increase dramatically and they would devour crops and grains at an alarming rate.

The use of pesticides destroys insects, but this also starves the birds. The predator birds dependent on small birds and rats for their food then reduce in number and the rat population will go unchecked.

If the predator animals are hunted down, the grazing animals multiply in large numbers and will eat up all the green grass, leaving the land barren and unproductive.

In contrast, if the predators were in excess of the grazing animals they would overrun the land and man, himself short of food, will become their prey.

When the ocean currents become unusually cold, plankton do not survive and this ultimately leads to the death of seals - the staple food of the Eskimos.

An important conclusion that can be drawn on studying the food chain is that the amount of green plants in any environment should be much more than the animals or insects that feed on them and they in turn should be in excess of the predator species. This is the origin of the famous Chinese saying that "only one tiger can live on a hill". In the ecological pyramid only one consumer is shown at the apex while the broad base is filled with the green plants.

Thus the importance of each and every creature in the web of life is evident. Tampering with the food chain only produces negative results, leading to the destruction of species. This shows why biodiversity and all its components are essential to maintain the balance of nature. Man is only a strand in the delicate web of relationship in the global ecosystem. Every time a species becomes extinct, the strand is broken and man himself moves closer to his doom.

The seriousness of the problem is multiplied enormously when the plants get extinct, as they form the base of the ecological pyramid constituting the "primary producer" of the food chain upon which depend all other species of life including human beings. Loss of plants breeds desertification and loss of soil wealth.

Our own species is subject to evolutiionary processes just as any other life form and the more we learn about other species, the better will we be equipped to manage ourselves.

The greater the diversity of wild species the greater is the opportunity for the production of new varieties. Almost all crops and livestock on which we depend for food production are hybrids - the result of cross breeding - originally from wild species. Compared to their wild ancestors, hybrid species are vulnerable to an epidemic or any abnormal climatic variation, which may completely wipe them out. Without their wild ancestors, it might not be possible to replace them.

The medicinal qualities of a variety of plant species are yet to be unfolded. The greater the variety of wildlife we conserve, the more will we be able to learn from it.

Causes for Biodiversity Loss

Today, we are rapidly eliminating the conditions necessary for the continued existence of biodiversity due to overexploitation and ignorance. These losses have now reached critical levels and are beginning to be addressed as biodiversity loss. Mankind has altered the environment for millennia, causing some species to prosper and others to suffer. Human activity is now changing the natural environment at a rate unprecedented in evolutionary history, and mass extinctions are a certainity.

Loss of biodiversity is a threat to civilization second only to thermo-nuclear war in its severity. The consequences could be quite incalculable. All other environmental problems like pollution, global warming and ozone depletion can be overcome but not erosion of biodiversity or extinction of species. Species once lost from the face of the earth cannot be brought back.

Major Causes for Biodiversity Loss

Loss of biodiversity occurs when either the habitat essential for the survival of a species is destroyed, or particular species are destroyed. The former is more common as habitat destruction is a fallout of development. The latter reason is encountered when particular species are exploited for economical gain or hunted for sport or food. The African elephants whose numbers have drastically declined are ruthlessly killed for their tusks, while the lions and tigers are being hunted for sport and the dodo, a bird native to Mauritius, was exploited for its tasty flesh while at the same time its habitat was destroyed and thus became extinct.

Extinction of species may also be due to environmental factors like ecological substitutions, biological factors and pathological causes which can be caused by nature or man.

Natural Causes for the Loss of Biodiversity

Natural causes include floods, earthquakes, landslides, natural competition between species, lack of pollination and diseases.

Man-made causes for the loss of biodiversity

Destruction of habitat in the wake of developmental activities like housing, agriculture, construction of dams, reservoirs, roads, railway tracks, etc.

According to Food and Agricultural Organisation (FAO), the closed tropical forests are disappearing at the rate of 7.5 mha per year and the open forests at the rate of 3.8 mha a year. According to UNEP, roughly 16-20 mha of tropical forests are being burnt or axed each year. i.e. more than the size of a football pitch every second for various developmental activities and for meeting the requirements of food, fodder and fuel for the growing human and livestock population.

The rate of deforestation in the tropics is much faster than in other parts of the world and is equivalent to about 21.5 ha/min. In India, about 1.5 mha of prime forests were felled during the 1980s. This has played a major role in the loss of biodiversity.

Pollution, a gift of the industrial revolution can be given the pride of place for driving a variety of species in air, water and land towards extinction.

Motorcars, air-conditioners and refrigerators, the three symbols of a modern, affluent society, have been instrumental in global warming and ozone depletion. They have drastically altered the climate with disastrous effects on the various species. Factories and power stations spewing out poisonous gases and effluents have fouled up the environment bringing death and disease to many species. Oil spills and discharge of sewage have ravaged the oceans and coastal habitats.

A large number of species are threatened by over hunting, poaching and illegal trade. In 1979, Africa had an estimated 1.3 million elephants. Less than 650,000 are left today.

Indiscriminate use of toxic chemicals and pesticides and overexploitation of wildlife resources for commercial purposes are responsible for the rapid decline in the number of some species. The tiger for instance is hunted for its claws and other parts believed to be effective cures for various ailments of man. Snakes and crocodiles are killed in large numbers for their skin and minks, sable, ermine, etc., are in demand for the luxury and warmth of their fur.

Genetic erosion arises from the loss (due to commercial and anthropogenic pressures) of habitats rich in biodiversity and from the disappearance of the traditional conservation practices of wild species in their habitats by rural and tribal people.

"We can marvel at the colours of a butterfly, the grace of a giraffe, the power of an elephant, the delicate structure of a diatom. Every time a species goes extinct, we are irreversibly impoverished".

Even the loss of a single species is a tragedy, because each form of life is a storehouse of irreplacable genetic resources. Extinction is an irreversible process and when a species becomes extinct it sets off a cascade of extinction.

Present Rate of Extinction of Species

In the past, extinctions were mainly effected by natural causes. But today man is the cause of most if not all extinctions. According to current estimates, on an average 900,000 vertebrate species have become extinct every 1 million years during the last 200 million years. This is roughly about 90 species every century. In the case of higher plants it is estimated that roughly 1 species has become extinct every 27 years.

In recent years oceanic islands have suffered mass extinctions due to destruction of lowland forests and introduction of predators, disease and aggressive, weedy plants by man. Since their area is small the population of species is also limited. It is easier for smaller numbers to be destroyed completely. Coral reefs are marine equivalents of rain forests and are home to a greater variety of species than any other in the marine environment; but, their submerged plight is less obvious than that of the rain forests. The movement of a trawler through them is the equivalent of a bulldozer through the rain forests. About 75% of birds and mammals that have become extinct in recent years were island dwelling species.

The rapid destruction of the tropics and other diverse ecosystems has given rise to a genuine fear that a quarter of the world's biodiversity is under risk of extinction in the next 20-30 years. Available data suggests, that one species has become extinct every four years between 1600 and 1900 A.D. But from 1900 onward, the extinction rate averages one per year. In the last four hundred years,

120 species of mammals and 225 species of birds have become extinct in the world.

IUCN Categories of Threat

All plant and animal species are in a state of dynamic equilibrium and form a closed web in a stable ecosystem. However, even under natural conditions, extinction of species is brought about by geological and evolutionary changes in time and space. Natural disasters like earthquakes, floods and forest fires are the cause of habitat loss. The loss of habitat has been accelerated due to the predominance of human beings on the land and the accompanying industrial revolution and population explosion. As a result, some of the most beautiful and interesting plant and animal species have been lost forever.

The accepted categories of existence or otherwise of plant and animal species is given below based on the International Union for Conservation of Nature and Natural Resources (IUCN).

Normal

Species whose population levels are considered to be normal for their survival.

Endangered

Species which are in danger of extinction. The survival of such species is difficult if the negative factors that have led to a decline in their population continue to operate. Endangered species include those whose numbers have been reduced to a critical level and those whose habitats are under threat.

Examples: Black buck, Crocodile, Wild ass, Indian rhino, Lion tailed macaque.

Vulnerable

Species whose population has declined to levels from where it is likely to move into the endangered category in the near future, if the negative factors continue to operate.

Examples: Blue sheep, Asiatic elephant, Gangetic dolphin.

Rare

Species with a small population that are at present neither endangered nor vulnerable but are at considerable risk. These species may move into the endangered or vulnerable category if the negative factors affecting them continue to operate.

Examples: Himalayan Brown Bear, Wild Asiatic Buffalo, Desert Fox and Hornbill.

Extinct

Those species which are not found after repeated searches of known or likely areas where they may occur. A species may be extinct from a local area, region, country, continent or the whole earth.

Examples: Asiatic Cheetah, Pink Headed Duck or Dodo.

Endemic

These are species which occur in a restricted area usually isolated by natural or geographical barriers.

Examples: Andaman Teal, Nicobar Pigeon and Andaman Wild Pig.

Animals in Danger

As wildlife become extinct or their population declines, the balance of nature is severely upset. The disappearance of a wild species can alter a food chain, upsetting the delicate balance between predator and prey and cause insect pests to multiply. Malaria is on the upsurge in Asia because of over harvesting of bull frogs which eat the mosquitoes spreading the disease.

Out of 350 species of mammals, 81 are greatly endangered. There are about 2100 species and sub species of birds which are endangered. Of the reptiles, the important ones are a species of tortoises, 5 species of monitor lizards and 3 species of crocodiles and turtles. Some snakes are hunted for their flesh, skin, hide, shell, teeth, bones etc.

Plants in Danger

Studies undertaken on flora in several parts of the world have shown that many plant species are in danger of extinction while some have become extinct recently. The International Union for Conservation of Nature and Natural Resources (IUCN) have estimated that about 10% of the world vascular plant species totalling about 20,000 to 25,000 species are under varying degrees of threat.

Endangered Abies pindrow, Drosera burmannivahi.

Vulnerable Colchicum S.P, Luceum S.P., Magnolia griffithil.

Rare Drosera burmanni, Gnetum ula.

Biosphere Reserves

The concept of biosphere reserves is the key to achieving the objectives of the MAB (Man and the Biosphere Programme) which aims at striking a balance between conserving biodiversity, encouraging economic and social development and preserving cultural values. Biosphere reserves are areas of terrestrial and marine ecosystems where, through appropriate management mechanisms, the conservation of ecosystems and their biodiversity is ensured. Biosphere reserves are designed to bring people and nature together to demonstrate how to both use and preserve nature.

In 1968, the UNESCO conference on the conservation and rational use of the biosphere within UNESCO was held. The primary goal of UNESCO-MAB is the exchange of scientific knowledge and management experience. There are currently more than 375 biosphere reserves in 85 countries, forming a global network of scientists and natural resource managers working to maintain the longterm survival of fragile ecosystems.

The biosphere can be defined in two different ways.

- From the perspective of the physical scientist, the biosphere is that part of the earth that contains living organisms, viz., the atmosphere, the crust of the earth and water.
- From the perspective of the ecologist, the biosphere is that part of the planetary ecosystem that is composed of living organisms. If we define all organisms as belonging to a single subsystem, that subsystem would be called Biosphere.

The biosphere encompasses all the life in air, water and soil.

Biosphere Reserves in India

The basic purpose of creating biosphere reserves is the conservation of the biotic community as a whole. So far we have 7 biosphere reserves and 7 other potential sites have been identified to be declared as biosphere reserves.

Established Biosphere Reserves in India

Name of the Reserve	*State*	*Area (sq.km.)*
Nokrek	Meghalaya	50
Manas	Assam	600
Great Nicobar	Andaman & Nicobar	1,045
Nandadevi	Uttar Pradesh	2,000
Similipal	Orissa	2,750
Nilgiri	Karnataka, Kerala and Tamil Nadu	5,520
Sunderbans	West Bengal	10,330
Gulf of Mannar	Tamil Nadum	10,500

Proposed Biosphere Reserves in India

Name of the Reserve	*State*	*Area (sq.km.)*
Kaziranga	Assam	760
Kanha	Madhya Pradesh	940
North Andamans	Andaman & Nicobar	1,376
Namdapha	Arunachal Pradesh	2,500
Thar Desert	Rajasthan	3,162
Uttarkand	Uttar Pradesh	3,940
Rann of Kutch	Gujarat	5,000

Biosphere Reserves in South India

The Nilgiri Biosphere Reserve has been formed as per the guidelines of the UNESCO in tropical, terrestrial environmental ecosystem on 23.9.1986. The biosphere reserve extends over Karnataka, Kerala and Tamil Nadu in South India, and encompasses an area of approximately 5670 sq. km. The Nilgiri Biosphere Reserve covers certain forest areas in the Nilgiris, Periyar and Coimbatore districts. It includes the forested slopes of Nilambur and Nilgiri, the upper Nilgiri plateau, the Silent Valley and Siruvani Hills.

Flora and Fauna

The forest types found in the Nilgiris are, thorn scrub forest, dry deciduous forest, moist deciduous forest, wet evergreen forest, montane forest and grasslands (sholas) and marshes and swamps. Woody climbers and epiphytes are also found. Some of the common trees found here are sandal, Indian rosewood, bijasal, jackfruit, jamun, ironwood, rhododendron, hill gooseberry, etc. Many varieties of orchids and the unique kurinji shrub that produces a blue coloured flower to which the Blue Mountains owe their name are also found here. Climbers like the black pepper and numerous herbs thrive here.

Animals like the elephant, bison, lion tailed macaque, sloth bear, pangolin, Nilgiri tahr, etc., birds like the great Indian hornbill, Nilgiri pipet, Nilgiri wood pigeon, Nilgiri flycatcher, etc., and numerous snakes and invertebrates are common here.

Gulf of Mannar Biosphere Reserve

The Gulf of Mannar Biosphere Reserve is situated along the southern coast of Tamil Nadu. The biosphere reserve includes the Gulf, the adjoining coasts and also the small islands dotting the gulf. The reserve was set up on 18th February 1989 and covers an area of about 10,500 sq.km. The reserve also includes a Marine National Park.

The Gulf of Mannar with its islands provides a unique habitat for the living organisms, that are in close intimacy with water.

Flora and Fauna

About 160 species of algae have been recorded here of which some 30 species are edible sea weeds. The area is also rich in sea grasses which provide food for sea mammals, particularly the dugong. The mangrove vegetation of the islands consists of species of rhizophora, avicennia, brugeira and ceriops.

The Gulf area has beautiful coral reefs that harbour a wide variety of marine vegetation and animals. Productive beds of pearl oysters, prawn species, edible bivalves, sea anemones, ascidarians and the sea cow Dugong dugon occur in the Gulf and the Palk Bay.

Mammals like langur, jungle cat, smooth Indian otter, etc., birds like grey billed pelican, black ibis, spotted crane, Indian river tern, etc., reptiles such as crocodile, starred tortoise, sand boa, etc., and fishes such as salmon, heckel, etc., are found along the island.

Marine Biodiversity

The long and complex process of evolution of life began below the surface of the sea. Today oceans form the largest environment for living organisms on earth. A huge variety of living organisms inhabit every part of the ocean - from the surface water to the deepest depth. The inhabitants range from the billions of microscopic creatures that thrive on the surface water to the gigantic blue whales. The forms of animals and plant life in the waters are incredible in their biological diversity.

The ocean provides shelter to different kinds of life forms, thereby acting as a greater reservoir of marine biodiversity. By year 2000, 60% of the world's population will be living or depending on the marine ecosystem.

Marine biodiversity includes 28 phyla of animals and 13 phyla of endemic animals. If plants and protista are considered, 88% of all phyla are found to be marine. Coral reefs and the mangrove vegetation zone are known as the most diverse communities in the seas. They include micro organisms such as algae, fungi, bacteria, protozoa, viroids and viruses, many species of mammals, fish, sponges and crustaceans. These animals form a place of high biological diversity. Therefore the world's biological diversity greatly depends on the ocean.

Until very recently, it was considered that man could not change the seas and their biological diversity. In the ocean, changes are taking place similar to those on land and much more rapidly. In the coastal zone, significant consequences can be expected of the Green House Effect and while human activities continue to extend beyond the shelf-edge, even those deep-sea reservoirs of biological diversity may be at risk.

On the other hand, marine ecosystems are most vulnerable to large-scale environmental changes. The oceans absorb all our waste and this alters the marine biological diversity, as they do not have internal adaptability. Proper conservation methods and sustainable management should be inculcated to restore this fast-changing habitat with its unique biodiversity.

Mangrove Ecosystem

Mangroves are salt-tolerant forest ecosystems of tropical and sub-tropical regions of the world. Along the sandy and rocky shores of estuaries and muddy sea coast, in places where any other tree would perish as a seedling, beneath daily tides and a salty diet, live and grow a special type of trees called the Mangrove vegetation. Mangroves anchor themselves to the soil by sending out long roots from trunks and branches. These act as stilts to support the tree and prevent it from toppling over. The muddy water where they live contains very little oxygen, but mangroves have an ingenious solution to this problem too. The trees send out a second air-breathing root system.

Mangroves are the spawning grounds, nurseries and feeding grounds for economically important fishes and crustaceans. They act as a buffer zone and offer protection to vulnerable communities such as coral reefs. They also stabilize bottom sediments, control local mean water level and the direction of flow. The mangrove ecosystem acts as a shield against cyclones and high tide winds. The mangrove ecosystem acts as a permanent reservoir of biodiversity wherein it supports many fishes, shrimps, crabs, trees, birds, etc.

Recently, mangroves are exploited directly or indirectly for food, fodder, fuel, timber, medicines and fisheries. It is estimated that during the last thirty years, almost 28% of the mangrove regions in India have been reclaimed for agricultural purposes, industrialisation and urbanisation. This is causing serious ecological imbalances in the coastal environment. Aquaculture is also disturbing and displacing this unique ecosystem.

Culture and Biodiversity

Despite all threats, species diversity and diversity within species still thrive. Their continued existence is due to farmers and others like the several tribal communities living within the forest. Their cultural practices and knowledge systems have helped to nurture biodiversity.

Traditional beliefs and practices have helped to preserve flora and fauna - in certain areas - in their pristine glory, as a great ecological heritage for modern civilisation. Traditional practices still followed by the tribals do not commercialize forest products and never exploit them in an unsustainable manner.

The universe along with its creatures belongs to the Lord. No creature is superior to any other. Human beings should not be above nature. Let no one species encroach over the rights and privileges of other species.

Isha-Upanishads (1500-600 B.C.)

Sacred groves - a wealth of diverse and rare species

Nature worship is a tribal belief based on the premise that all creations of nature have to be protected. Such beliefs have preserved several virgin forests in pristine form called Sacred Groves (the forests of God and Goddesses). These patches of forest or parts of large forests have been left untouched by the local people and any interference with them is banned. The practice dates back to about 3000 to 5000 B.C.

Certain societies revere a particular tree which they have preserved from time immemorial. The Mundas and the Santhal of Bihar worship mahua (Bassia latifolia) and kadamba (Anthocaphalus cadamba) trees and the tribals of Orissa and Bihar worship the tamarind (Tamarindus indica) and mango (Mangifera indica) trees during weddings. Many a banyan tree is considered as sacred in India.

Indian society comprises of several cultures, each with its own set of traditional methods of conserving nature and its creations. Sacred qualities are often ascribed to springs, mountain peaks, plants and animals which are closely protected. Troops of macaques and langurs are found around temples. Around Bishnoi villages in Rajasthan, herds of blackbuck, chinkara and nilgai roam.

In densely populated Bangladesh, the only surviving population of the turtle species, Trionyx nigricans, lives in a pond sacred to a Muslim Saint called Byazid Bostan.

Biodiversity Conservation

Having been through the reasons for the need for biodiversity and the effects of loss of biodiversity it becomes evident that ultimately the survival of the human race itself is threatened. Therefore the conservation of biological diversity is of paramount importance to the survival of man and only he can initiate and sustain a strategy for the same.

The main objectives and advantages of biodiversity conservation are as follows:

- Conservation of biological diversity leads to conservation of essential ecological diversity to preserve the continuity of food chains.
- The genetic diversity of plants and animals is preserved.
- It ensures the sustainable utilisation of life support systems on earth.
- It provides a vast knowledge of potential use to the scientific community.
- A reservoir of wild animals and plants is preserved, thus enabling them to be introduced, if need be, in the surrounding areas.
- Biological diversity provides immediate benefits to the society such as recreation and tourism.
- Biodiversity conservation serves as an insurance policy for the future.

The Conservation practices should include:

- Policy shifta and integrated land use
- Policy amendments are the necessary first step towards conservation. National policies dealing directly with forestry or influencing biodiversity indirectly through land tenure, rural development, family planning or subsidies for pesticides or energy can have significant impacts on the conservation of biodiversity. For e.g., removing subsidies for forest clearing in Brazil is a powerful response to deforestation.

Four major international moves are afoot to conserve biodiversity.

Owing to the need for more conservation funding in developing countries, the World Bank and the United Nations Development

Programme (UNED) have established the Global Environmental Facility in 1990 on a three year pilot basis.

The international biodiversity strategy programme organised by the World Resources Institute, the World Conservation Union and UNEP involving more than 40 governmental and nongovernmental organisations from around the world released the global biodiversity strategy in February, 1992.

Under UNEP's auspices, more than 150 nations worked to negotiate a "Convention on Biological diversity" at the "Earth Summit", in Rio de Janeiro, Brazil in 1992. This convention seeks to establish a legal framework governing international financial support for biodiversity conservation, the identification of international conservation priorities, and the transfer of technologies for both the conservation and the use of biodiversity.

The Earth Summit produced Agenda 21, a plan of action on a number of issues, including biodiversity conservation in the 21st century.

Ex situ Conservation

Conserving biodiversity outside the areas where they naturally occur is known as ex situ conservation. Here, animals and plants are reared or cultivated in areas like zoological or botanical parks.

Reintroduction of an animal or plant into the habitat from where it has become extinct is another form of ex situ conservation. For example, the Gangetic gharial has been reintroduced in the rivers of Uttar Pradesh, Madhya Pradesh and Rajasthan where it had become extinct.

Seedbanks, botanical, horticultural and recreational gardens are important centres for ex situ conservation.

In situ Conservation

Conserving the animals and plants in their natural habitats is known as in situ conservation. This includes the establishment of:

- National parks and sanctuaries
- Biosphere reserves
- Nature reserves
- Reserved and protected forests

- Preservation plots
- Reserved forests
- Agrobiodiversity conservation

After the introduction of cotton, tobacco, sugarcane, sunflower, soyabean and so on, farmers became victims of monocultures in their greed for money. Therefore many of the indigenous varieties of crops were lost. Moreover, the hybrid varieties of fruits and vegetables (e.g. tomatoes), introduced for pulp are more susceptible to disease and pests. Though hybrid varieties are preferred, traditional wild varieties of the seeds should be conserved for future use in the event of an epidemic which would completely wipe out the hybrids.

Botanical gardens, agricultural departments, seed banks etc., alone should not be given the responsibility of agrobiodiversity conservation. Every farmer, gardener an cultivator should be aware of his role in preserving and conserving agrobiodiversity.

Conserved areas of Tamil Nadu

Wildlife sanctuaries	*Area in hectares*
Indira Gandhi Wildlife Sanctuary	96,003
Mundanthurai Wildlife Sanctuary	56,700
Srivilliputur Grizzled Squirrel Wildlife Sanctuary	48,520
Mudumalai Wildlife Sanctuary & National Park	39,946
Kalakad Wildlife Sanctuary	25,064
Mukurthi National Park	7,846
Point Calimere Wildlife Sanctuary	1,726
Vallanadu Blackbuck Wildlife Sanctuary	1,641
Gulf of Mannar National Park	623
Arignar Anna Zoological Park	510
Arignar Anna Zoological Park	510
Guindy National Park	300
Bird sanctuaries	
Pulicat Lake Bird Sanctuary	15,367
Vaduvur Bird Sanctuary	128
Kanjirangkulam Waterbird Sanctuary	104
Sithirangudi Waterbird Sanctuary	48
Udhaya Marthandapuram Bird Sanctuary	45
Vettangudipatti Waterbird Sanctuary	38

- Biosphere Reserves
  Gulf of Mannar Biosphere Reserve
  Nilgiri Biosphere Reserve

Categories of Forests in Tamil Nadu

Reserved Forest	1,948,600 hectares
Protected Forest	252,800 hectares
Unclassified Forest	61,400 hectares
Total area under forest cover	2,262,800 hectares

Conserved Areas of India

The biodiversity in India is being conserved through the establishment of Sanctuaries, National Parks , Biosphere Reserves , Reserved Forests, etc., after the Wildlife Protection Act, 1972. But even before this Act had been passed, some states had their own legislations and the Indian Forest Act of 1927 allowed the setting up of wildlife sanctuaries.

Some of the important wildlife regulation acts are:

- Elephant Preservation Act, 1879
- The Indian Fisheries Act, 1897
- Wild Birds and Wild Animals Protection Act, 1912
- The Indian Forest Act, 1927
- Bengal Rhinoceros Act, 1932
- Haily National Park Act, 1936
- Bombay Wild Animals and Wild Birds Protection Act, 1951
- The Cruelty Against Animals Act 1960
- The Wild Life (Protection) Act, 1972
- The Forest (Conservation) Act, 1980
- Wildlife (Protection) Amendment Act, 1991
- The Conservation of Forest and Natural Ecosystems Act, 1994

National Parks and Sanctuaries

These have been created exclusively for protecting the wild flora and fauna in all parts of India and for the purpose of protecting and propagating wildlife.

Biosphere Reserves

Their basic purpose is the conservation of the biotic community as a whole. So far we have 7 biosphere reserves and 7 other potential sites have been identified to be declared as biosphere reserves.

Botanical Gardens

Their main objective is to study and conserve the diversity of plant life of a region or nation. This is known as ex-situ conservation under which rare and threatened plants may be raised in a botanical garden so as to conserve them for posterity. There are about 78 well established gardens in India.

Zoological Parks

Zoos or Zoological Parks are important centres for the ex-situ conservation of wild animals, particularly those which are rare and threatened with extinction. There are about 109 zoological parks, 53 deer parks, 13 safari parks, 7 snake parks, 26 breeding or education centres and 7 aquariums presently working for conservation.

Projects to Save Threatended Species

Project Tiger

Project Tiger was initiated as a Central Sector Scheme in 1973 with 9 tiger reserves located in different habitat types in 9 different states. There are totally 18 Reserves in 13 states. At present tiger Conservation has been viewed in India not only as an effort to save an endangered species but, with equal importance, also as a means of preserving biotypes of sizeable magnitude.

Crocodile Breeding Project

The project was started in Orissa and then extended to several other states in April 1975 with UNDP assistance. The main objective was to protect the three endangered species of crocodiles namely - Gavialis gangeticus, Crocodylus palustris and the salt water crocodile, Crocodylus porosus.

Lesser Cats Project

The project was launched in 1976 with the assistance of WWF in India for conservation of four species of lesser cats e.g. Felis bengalensis Kerr, Felis marmorta Martin, Felis lemruinki Vigors Horsfield and Felis viverrina Bennet, found in Sikkim and Northern part of West Bengal.

The Manipur Brow-Antlered Deer Project

This was launched in 1981 in Manipur to save the brow-antlered deer (Cerevus eldi eldi) which is on the verge of extinction. The habitat includes 35 sq.km. of park and sanctuary. The population of the deer has increased from 18 to 27.

Project Elephant

It was launched in 1991 to protect the Asiatic elephant which is also a highly endangered species because of large scale poaching.

Project Rhino

It was launched in 1987 in Kaziranga Wildlife Sanctuary in Assam to save the lesser one horned rhinoceros from extinction. It covers an area of 430 sq.km. and is the natural home of the dwindling rhino.

Class	*Threatened*	*Extinct*
Mammalia	53	2
Aves	69	2
Reptiles	23	-
Amphibia	3	-
Invertebrates	22	-
Total	170	4

Category of threat	*Number of species*
Endangered	24
Rare	45
Vulnerable	49
Indeter[illegible]	28
Insufficienty known	24
Extinct	4

Himalayan Musk Deer Project

This was launched in 1981 to save the endangered musk deer which is facing extinction. Captive breeding has yielded good results.

Project Hangul

This project was launched in 1970 in Kashmir valley to save the highly endangered Kashmir stag (Cerevus elaphus hanglu) which is facing extinction. As a result their population has increased to about 500 now.

Conservation - A People's Movement

If all the beasts were gone, we would die from a great loneliness of spirit... All things are connected. Whatever befalls the earth, befalls the children of the earth.

Chief Seattle, 1854

Protecting the environment is everyone's responsibility. There is an urgent need for awareness about the devastation of the environment brought about by man's greed in a materialistic era. The swelling population of the world is insatiable in its demand for food, water, air and land. As more forests are cleared, the biodiversity they were home to is lost forever. About the middle of the 21st century we will see the last of the primary tropical forests and the extinction of over half of the 1.4 million named species. The poorer but more populous nations will scrape up every available resource in their losing battle for survival. The never ending needs of man will make deep inroads into natural resources and will undoubtedly take him closer to extinction with each day. Only such of those species which have survived for several million centuries like the cockroach, will survive the catastrophe.

There is an increased awareness among the people towards the conservation of ecologically sensitive areas. Greenpeace - a group devoted to environmental protection - was responsible for helping to ban whaling. Closer home, the Chipko movement initiated by Sunderlal Bahuguna stopped felling of trees in some parts of the Himalayas. The Silent Valley which would have been flooded had the proposed dam been built was also saved by the intervention of environmental activists and groups. In our own small way each of us can contribute to conserving biodiversity in one form or the other. Some of the activities in the areas of environmental conservation include :

- Holding rallies and marches to bring about a mass awakening to environmental issues like deforestation, construction of dams or factories in environmentally sensitive areas like the Silent Valley, etc.
- Conducting action oriented programmes like social forestry, one house, one tree project, afforestation projects, etc.
- Spreading awareness through mass media like television, radio and publications.
- Creating awareness through environmental training programmes, folkart media like street plays, etc.
- Introducing environmental legislation to tackle the sensitive environmental issues.
- Mass movements like Chipko, Narmada bachao andolan, etc.

Methods of Saving Biodiversity

Every individual can make a small though significant effort in the race to save our planet and conserve biodiversity. Man is only yet another of nature's creatures and not an alien to the other life-forms. He has no more right to destroy nature than any other being. We should treat all animals and plants with compassion.

"Man does not have the right to destroy what he cannot create. The human being is not an alien to exploit it".

Convention of Biological Diversity

The Convention on Biological Diversity was signed by 156 nations and the EC in Rio de Janeiro, June 1992. For the convention to come into force, 30 countries need to ratify it. The aim of the convention is to save species and plants from extinction and their habitats from destruction.

The developed countries are looking for a sustainable supply of biological resources from the developing countries and easy access to them as well. The developing countries lacking the technology to exploit their resources are inviting the developed countries to do so. This has resulted in the developed nations channelling out the benefits of these natural resources. The developing countries are now demanding a higher share of the accrued economic benefits. The developed nations are also concerned by the unsustainable exploitation of natural wealth, particularly rainforests.

Key Points from the Convention on Biological Diversity

The aim of the Convention on Biological Diversity is 'the conservation of biological diversity, the sustainable use of its components and the fair and equitable sharing of the benefits arising out of the utilization of genetic resources. The convention stipulates that Parties must :

- develop national strategies for the conservation and sustainable use of biological resources;
- establish protected areas, restore degraded ecosystems, control alien species, and establish ex-situ conservation facilities;
- establish training and research programmes for the conservation and sustainable use of biodiversity and support such programmes in developing countries;
- promote public education and awareness of the conservation and sustainable use of biodiversity; recognize the right of governments to regulate access to their own genetic resources, and, wherever possible, grant other Parties access to genetic resources for environmentally sound uses; encourage technology and biotechnology transfer particularly to developing countries; establish an information exchange between the parties on all subjects relevant to biodiversity; promote technical and scientific cooperation between parties (particularly to developing countries) to enable them to implement the convention.

3

Forests and Forestery

Introduction

- A dense forest is like an umbrella. The trees provide shade and prevent the soil from drying up, thereby maintaining the moisture content of the soil.
- Forests lower the temperature and reduce sunlight.
- The increased humidity results in the formation of fog and dew in nearby cultivated land.
- Forests reduce the velocity of wind and rainfall, thereby preventing soil erosion.
- The dry leaves and other organic matter from the trees get decomposed and mix with the soil to form humus which is fertile and can absorb large quantities of water. Thus the soil under the forest becomes a giant sponge which absorbs water during the rainy season for the hot summer months.

Forests and Rainfall

- The water from the soil is absorbed by the tree.
- This water is released into the atmosphere through the process of transpiration, thereby increasing the moisture content of the atmosphere and bringing rainfall.
- Deforestation could lead to changes in surface conditions, which would increase the intensity and decrease the duration

of rainfall, thereby increasing run-off. This causes soil erosion, leading to silting of river beds. This is how floods occur.

- Floods cause a lot of damage to the standing crops and there can be drought even after a heavy rain. This is because the number of rainy days is reduced, leading to drought conditions and water scarcity.
- When drought occurs, dust is formed, leading to desertification, because the dust particles lead to dessication which dries the land.

Important Facts

- 50 acres of tropical rain forests are cut down every minute all over the world.
- There will be no rain forests left 20 years from now.
- The world's oxygen balance depends on trees and forests.
- According to India's National Forest Policy (1952), 33% of the land should be under forest cover, but, at present, the area under forest cover in India is only 10%.
- India is losing 1.5 million hectares (mha) of forests per year, thereby bringing down the total forest area from 74 mha to 40 mha.
- Deforestation causes loss of top soil to the tune of 12,000 million tonnes.
- Due to deforestation, India loses Rs.10,000 crores every year in the form of damage by floods.
- Large-scale felling of trees causes irreparable loss of organic resources.
- India requires 80 million m3 of fuel wood every year.
- The Government of India promulgated the Forest (Conservation) Act in 1980 to check indiscriminate deforestation/diversion of forest land to non-forest purposes.

Photosynthesis

Carbohydrates are synthesized from carbon dioxide and water by utilizing the radiant energy of light, which is captured by the chlorophyll in the cells of the green leaves. Photochemical reactions are initiated by which oxygen is released from water and light energy is converted to chemical energy, and amino acids are synthesized by the combination of intermediate products with elements such as nitrogen, derived from mineral salts. Most of the oxygen in the atmosphere is derived from photosynthesis. This makes trees and plants essential for our survival.

Forests and the Greenhouse Effect

- The burning-up of fossil fuels releases a lot of carbon dioxide (CO_2) into the atmosphere. This CO_2 acts as a one-way shield, which allows the sunlight to pass through it and is reflected by the surface of the earth back to the atmosphere.
- The CO_2 shield does not allow this reflected light to escape and, thereby, warms up the atmosphere, leading to an increase in the global temperature.
- This increase in temperature may cause the melting of the polar ice-caps, leading to a rise in the ocean levels, which can be catastrophic.
- Forests and trees can greatly reduce this danger. As all of us know, trees absorb CO_2 from the atmosphere during the day time, use it for photosynthesis and give out oxygen. If the number of trees and forests increase, the CO_2 levels will come down. Thus the Green House Effect can be mitigated.

Afforestation

The prosperity and economic growth of a country depends, to a large extent, on its forest wealth. The forest is an important natural resource for any country and deforestation retards a country's development. To meet the demands of the expanding population, basic resources can be obtained only through afforestation. Nurseries play an important role in increasing forest cover. As necessary as it is for a child to get through her/his tender age at a nursery, so is it for plants to grow under proper care and protection. This prepares them to withstand adverse situations during planting.

There are greater possibilities of plants getting destroyed by natural factors such as storms and wind, grazing by cattle, etc., during direct sowing. Eventually, not only the seeds but our efforts also are wasted. Hence a successful afforestation programme can be achieved only through proper initial care.

Setting Up A Nursery

The proper growth of a tree depends on the quality of the seeds and the nursery techniques that are implemented. Therefore it is advisable to follow certain guidelines while raising nurseries.

Developing nurseries is not a difficult occupation, though it needs proper selection, care and maintenance.

Selection of site

The first necessary ingredient for nursery raising is the selection of site. The site can be near the plantation area to facilitate easy transportation at the time of planting. However, the basic facilities that are to be looked into are good sunlight, enough water resources and land (smooth and not undulating or rocky).

Nature of soil

If the soil is found to be fertile, there is no necessity to buy manure from outside. Otherwise soil should be prepared by mixing manure, sand and soil together.

Fencing

Since it is important to protect tender plants from being browsed by cattle or stolen, a proper fencing method should be adapted such as live fencing (the cheapest), stone wall, barbed wire, etc.

Implements

Spade, basin, sieve, sprinkler, sickle, scissors, etc. are the prerequisites for starting the nurseries.

Seed collection and storage

Seeds can be obtained from the nearest forest department or seed bank. If the nursery is a considerably large one, seeds can be collected by the raisers and stored.

Collection of seeds

Seeds should be collected only from a healthy and properly grown tree. Do not collect seeds from very young or very old trees.

Selection of seeds

While selecting seeds, care should be taken to avoid mutilated, broken or pest-affected seeds as they will not germinate, eventually leading to waste of time, money and man power.

After selection, seeds having thick seed coats should be broken and be partially exposed to facilitate easy germination.

If the seeds are not sown immediately they should be dried in open sunlight, packed and stored in air tight containers, free from pest. If the seeds or the containers are moist, there will be fungal growth.

Sowing

If the seeds are to be sown soon after collection, proper seed treatment is necessary. Some seeds which are tender and soft can be sown without prior treatment. But where seeds are strong and thick they need to be soaked in either cold water or in hot water for not more than 3 minutes or bathed in dilute sulphuric acid. Such treatment will help the seeds to germinate at a faster rate.

Preparation of bed

The preparation of the beds before sowing the seeds is important. The normal size of a bed is 1 - 1 1/2 metres and 2 to 4 metres in length. The bed should be made with a slight slope and flattened in order to facilitate proper drainage and infiltration. After the measurements are done, the bed has to be cleared of weeds, stones and pebbles.

If the soil is good, then the same soil can be used for the beds. Otherwise, good manure has to be added. Later, seeds can be sown. If the seeds are large, holes can be made with a small stick or pole on the bed and the seeds can be sown in a straight line.

If the seeds are very small, they can be sprayed over the bed and covered with soil.

After sowing the seeds, the bed can be covered with hay or coconut leaves to protect them from excess heat and pests (birds).

When the seeds germinate, a small roof made of coconut palm leaves can be constructed over the bed to prevent the tender plants from getting burned by direct sunlight.

Instead of watering directly, if water is sprinkled with a sprinkler, the seeds will not get dispersed or washed away.

Sowing in plastic bags

The normal size of plastic bags is 10 cm. width by 20 cm. length. However, the sizes will vary according to the species grown. Normally, a bag of 20 by 30 cms. (guage 250) is preferred.

Three holes should be made at the bottom of the bag to help the excess water to drain.

While filling up soil in the bag, the composition of the soil should be equal quantities of loam, sand and manure.

Before being filled in the bags, the soil should be dried and mixed well and filtered through a sieve to remove stones and pebbles and any other solid matter.

This fine mixture should be filled in airtight plastic bags. The bags should be arranged neatly in rows.

Two or three seeds should be sown in each bag, not very deep in the soil but the depth will vary according to their size. A name board should be kept for each species grown.

Once the seeds sprout, regular attention should be paid to check for weeds, pests and disease.

The polythene bags should be shifted regularly to avoid roots penetrating into the soil and the plants should be pruned periodically.

They should be arranged according to their height, to ensure adequate sunlight for all the plants.

While transplanting small saplings from beds to polybags, the bed should be watered so that the plants can be pulled up easily. The plant should be removed carefully without causing any damage to the roots. While transplanting them care should be taken that the main root does not bend and enough depth is provided for the plant.

Holding the stem or the tip of the plant while plucking for transplantation should be avoided. Instead, the middle part of the plant should be used as a grip to pluck it from the ground.

Plants can also be grown through grafting. A healthy branch of the selected species can be cut and the corners of the branch can be sliced off.

The sliced portion can be inserted into a soil filled plastic bag or seed bed.

Finally, while transporting the saplings from the nursery site to the plantation area, the plastic bags should be stacked properly in basins and transported, otherwise the plants could get damaged and die during transportation.

Planting

While planting the saplings the plastic bag should be cut vertically, on two sides and pulled apart. The sapling should be planted along with the soil.

Tree Planting and After Care

To regenerate wastelands, planting can be done using traditional techniques but it is more time consuming, labour intensive and economically not viable. A more modern and efficient method is transplanting saplings from the nursery to wastelands which are a few hundred hectares, for which some alternative techniques can be employed.

Before taking up this planting programme, the site should be surveyed and live fencing established if possible. Soil samples should be tested for the fertility status with special reference to nitrogen, phosphorus and potassium. Most tropical soils lack nitrogen. To supplement and enrich this, one can prepare the soil by sowing large quantities of Crotolaria and Medicago sativa (Alfa-Alfa) seeds (about 10 kg per hectare). These are termed nurse crops. This sowing can be done with the onset of the monsoon or the first shower. Allow these plants to grow for 3-4 months and plough if possible into the soil. Once this operation is completed, pits can be dug. The size of the pits should be the size of the saplings raised in the polythene bags. For example, saplings raised in a polythene bag of 30 by 20 cm. can be transferred into a pit of a similar size or slightly more than this. The lower half of the polythene bag can be cut and then transferred into the pit. This technique of plantation saves time and labour. The density of plant saplings per hectare varies from plant to plant. For instance, Subabul can be planted at a density of 10000 plants per hectare i.e., 1 plant per sq.m., whereas Tamarind can be planted at a rate of 1500 to 2000 plants per hectare.

Generally, to get an ideal tree cover, one can plant around 2000 plants per hectare.

The transplanting operation can be done during the monsoon, that is, October/November, and nature can be allowed to take care of the plants after that. If the transplanted saplings are kept away from cattle and interference from man, it is quite likely that the survival rate will be 70% to 80%.

Occasional watering can be done during the summer months. Weeding can also be avoided, if it does not hamper the growth of the plants. Pruning, thinning and replacement of dead plants can be done in the following year.

One of the other factors that ensures the survival rate is the age of saplings. Saplings more than six months or a year old, will definitely have more survival capacity than saplings which are just two or three months old. On lands with an undulating surface, planting can be done as shown in the figure below. This method would check soil erosion and allow more water infiltration.

A tree can be planted anywhere, in any place, provided the following basic amenities are available.

- Proper soil
- Priority of species
- Water facilities
- Maintenance or after care
- Marketing opportunities
- Maintenance

Each plot should be well laid out. Proper spacing of plants helps in good management of the orchard.

Different trees require difference spacing and pit sizes. The positions of pits are marked before digging begins.

Trees	*Spacing*	*Bit sizes*
Mango	30' x 30'	3' x 3' x 3'
Lemon	15' x 15'	2' x 2' x 2'
Guava	20' x 20'	3' x 3' x 3'
Sapodilla	25' x 25'	3' x 3' x 3'
Papaya	8' x 8'	2' x 2' x 2'

Trees	*Spacing*	*Bit sizes*
Pomegranate	15' x 15'	2' x 2' x 2'
Jackfruit	30' x 30'	3' x 3' x 3'
Indian Jujube	20' x 20'	2' x 2' x 2'

(These are common trees of the Indian plains.)

Digging of Pits

Pits for fruit trees are always large (size: 2' x 2' x 2' or 3' x 3' x 3').

While digging, different layers of soil are piled in different heaps. Only the better soil is used for filling.

Filling of pits: 1

Before filling, the pits are dusted with pesticides to keep out white ants. The pits are then filled, preferably with silt or top soil excavated from the pit or collected from the surrounding land.

Filling of pits: 2

The lower half of the pit is filled with good soil. The next nine inches are filled with a mixture of soil and a small amount of superphosphate (100 g). The next six inches have a layer of leaf litter and green manure. The top layer of six inches contains a mixture of soil, superphosphate (1 kg), neem or castor cake (1 kg) and farmyard manure (1 basket).

Planting method: 1

Dig out a pit large enough for the sapling. Planting is done preferably in the evening. Fill the pit with water and cut open the polybag, place the sapling in the pit without damaging the roots and fill the pit with soil.

Planting method: 2

The soil filled in the pit is to be compacted to remove any air pockets. The sapling is to be watered regularly. Loss of soil moisture by evaporation can be prevented by mulching. A black plastic sheet covering the sapling basin prevents evaporation of soil moisture and suppresses weed growth. Straw can also be used for mulching.

Aftercare: 1

First, the basin is shaped. As the tree grows, the basin is shaped into a ring and irrigation is done within the ring. The sapling is watered as per requirement. Regular maintenance of the basin is essential. The soil has to be loosened and weeded regularly.

Aftercare: 2

Regular application of fertilizer in appropriate doses is essential for good growth. Nitrogen and Potassium can be applied in the ring basin and mixed with the soil. Super phosphate is filled in holes about nine inches deep. Green manuring helps in improving the soil structure.

Aftercare: 3

The pruning of the lower branches helps in proper management of each plant. The species planted will depend on the quality of the soil. All these trees need proper maintenance and care. Since each has a unique function to perform, after care also needs to be done according to the purpose of the plantation.

Coppicing:

Individual trees are cut at the base between 15 to 75 cm. above ground level. New shoots develop from the stumps. When the sprouts develop if it is grown for pole production, only 2 to 3 sprouts need to be allowed to grow. Accordingly, for fuel too required sprouts can be allowed to grow. Several rotations of coppicing are usually possible with most species. *Eg.: Subabul, Acacia nilotica.*

Pollarding:

In this system, all the branches including the top of the tree, are removed at a height of approximately 2-3 metres above ground level. The main stem continues to increase in diameter, though not in height. The main stem can be cut for poles and an advantage of this is that new shoots are high enough from the ground to prevent them from being browsed by cattle. This can also be used for the management of living fences. Eg.: Azadirachta indica, Eucalyptus, Gliricidia sepium, Cassia siamia.

Lopping:

When wood is mainly used for fuel and leaves for fodder, this sytem of removing most of the branches is useful.

Pruning:

For maintaining fruit trees and trees for forage and for cropping and in fences, this system of pruning or removing smaller branches and stems is useful. Eg.: Subabul, Gliricidia Acacia.

Thinning:

This method is practiced to maintain desirable trees, by removing the poor and diseased trees and thereby reducing the competition for light and nutrients. This can be done according to the plant density.

Trees that can be grown in Marshy Areas

- Acacia auriculiformis : Babul
- Dendrocalmus strictus : Bamboo
- Derris indica : Pongamia

Trees that can be grown in Sandy Soils

- Anacardium occidentale : Cashew
- Casuarine equisetifolia : Casuarina
- Eucalyptus Sp. : Eucalyptus

Trees that can be grown in Acidic Soils

- Tamarindus indica : Tamarind
- Gliricidia sepium : Gliricidia
- Albizia procera : White Siris

Trees that can be grown in Alkaline and Saline Soils

- Acacia nilotica : Babul
- Azardirachta indica : Neem
- Cassia siamea : Kashod
- Leuchaena leucocephala : Subabul
- Prosophis juliflora : Vilayati Babul

Trees that can be grown in Clayey Soils

- Alibizia lebbeck : Siris
- Pithecellobium dulce : Madras Thorn
- Acacia nilotica : Babul
- Azadirachta indica : Neem

Tree Species for Wasteland Development

Acacia nilotica

Common Name: Babul

Soil requirements: Grows in arid, low rainfall areas having long dry seasons. Thrives on a wide range of soils from heavy clayey to rocky soils.

Uses: It is a high calorific value food, it is useful for fuel, to make farm equipment, and in timber and pulp production. The foliage and pods are good food for livestock.

Leucaena leucocephala

Common Name: White Babul

Soil requirements: Grows in high pH soils and limestone and rocky soils but cannot tolerate high acidic or water-logged soils.

Uses: Salvador type is the fast growing one. The wood is good for fuel, pole, timber and furniture. Foliage is used as fodder for livestock. It controls soil erosion and is useful in reforesting degraded pastures.

Prosopis juliflora

Common Name: Mesquite

Soil requirements: Grows in all kinds of soils.

Uses: Useful as fuel, light timber and in fencing; pods are a source of food for livestock. Foliage is not browsed.

Casuarina equisetifolia

Common Name: Casuarina

Soil requirements: Adapted to a wide range of soils including calcareous, saline and seashore areas.

Uses: Good for fuel, poles, timber and as raw material for pulp. Foliage is not browsed by livestock. It controls soil erosion, acts as a wind breaker and can withstand a polluted atmosphere.

Albizia lebbek

Common Name: Siris

Soil requirements: It requires well drained soil, and the trees grow well near sea shores and on dry, alkaline soils.

Uses: It is useful for fuel, furniture and timber. The bark is used in tanning. It has high medicinal properties. The foliage is browsed on by animals.

Artocarpus heterophyllus

Common Name: Jack Fruit

Soil requirements: Requires deep soils, but adapted to alkaline, acidic, laterite, sandy and clayey soils.

Uses : Useful for furniture, foliage browsed by livestock, fruits are edible.

Melia azadirach

Common Name: Persian Lilac or China Berry

Soil requirements: Grows in saline, alkaline and slightly acidic soils.

Uses: Good for fuel, timber and paper production. Foliage is browsed by livestock and it is good for agro-forestry.

Samanea saman

Common Name: Rain Tree

Soil requirements: Grows on poor soils, alkaline conditions. Not in water-logged areas.

Uses: Good as fuel wood and for furniture making. Foliage is browsed. Pods make nutritious food for livestock.

Thespesia populnea

Common Name: Portia Tree

Soil requirements: Grows in sandy, laterite, alluvial black clay, marshy and acidic soils and on land flooded by salt water.

Uses: The trunk is used in making bullock carts, central bars, match sticks, etc. Foliage is a good fodder.

Gliricidia sepium

Common Name: Mexican Lilac

Soil requirements: Grows well in all kinds of soils like dry, moist, eroded, sandy and lime stone soils.

Uses: Good for fuel and to make furniture and tools. Foliage is browsed by animals. This is good for live fencing and for control of soil erosion.

Bambusa arundinacea

Common Name: Bamboo

Soil requirements: Grows in deep, well drained, sandy loams, but can be grown in clayey, acidic and alkaline soils too. Does not grow well on shallow and gravelly soils.

Uses : Used for constructing houses, basket making and for making quality paper. Foliage is a fodder for livestock.

Terminalia arjuna

Common Name: Arjun

Soil requirements: Grows in humid, tropical conditions, cannot withstand drought, but grows well in a variety of soils like saline, alkaline, gravelly, etc.

Uses: Leaves are a good fodder for livestock. The wood is good for fuel and to make furniture.

Tectona grandis

Common Name: Teak

Soil requirements: Grows on a variety of soils, laterite, gravelly, acidic and alkaline.

Uses: Wood is used for timber and is highly durable.

Moringa oleifera

Common Name: *Drumstick Tree*

Soil requirements: Grows in all soil conditions but not in water-logged soils.

Uses: Eaten as a vegetable, good species for agro-forestry and for control of soil erosion.

Azadirachta Indica

Common Name: Neem

Soil requirements: It is suitable for areas with long, dry seasons. It grows on a wide range of soils including dry, clayey, shallow, rocky and sandy soils. It does not grow in water-logged, saline or deep, dry areas.

Uses: Resistant to pests, good for fuel and timber and to make furniture. The oil is used for soap. The cake is used for manure and to control soil-borne pests.

Mangifera indica

Common Name: Mango

Soil requirements: Grows on a variety of soils but prefers deep, well drained soils.

Uses: Good for fuel, timber and medium quality furniture. Foliage is browsed by livestock. Fruits provide a substantial income.

Ziziphus jujuba

Common Name: Indian Jujube

Soil requirements: Grows in a variety of soils from sandy, gravelly to clayey, saline, alkaline and dry soils.

Uses: Wood is good for fuel. The wood is also used for making agricultural implements. Fruits are edible and the leaves are browsed by livestock.

Tamrindus indica

Common Name: Tamarind

Soil requirements: Grows in a variety of soils from gravelly, sandy and deep clayey to acidic and alkaline soils.

Uses: Wood is heavy and is excellent timber. The fruits are edible. The pulp is used in cooking. Foliage is browsed by livestock.

Emblica officinalis

Common Name: Gooseberry

Soil requirements: Grows in all kinds of soil.

Uses: Good fuel, foliage is browsed by livestock. Fruits are rich in Vitamin C and are edible.

Syzygium cumini

Common Name: Jamun

Soil requirements: Grows in sandy, limestone and alkaline soils. A full-grown tree can withstand drought and flood conditions.

Uses: Wood is used for fuel. The bark contains tannin. Fruits are edible and the foliage is browsed by livestock.

4

Environmental Laws

Introduction

Every individual is expected to know the law. A fundamental principle in jurisprudence is that "ignorance of law is not excusable." However, the fact remains that most people are unaware of the law and of their legal rights. Often, people play the role of mute spectators to many serious problems that threaten their peaceful lives primarily because they are ignorant of their rights under the law. One such problem is the various types of environmental degradation which every citizen faces, such as the exhaust from automobiles, the obnoxious smells from open drains, loud music and other noise, polluted drinking water, and so on. These problems can be solved with the aid of specific laws that have been formulated for the purpose.

The need to incorporate environmental legislation into the training programmes resulted in the organisation of awareness - raising seminars on "Environmental Protection, People and the Law" for NGOs, lawyers and law students.

According to the Ministry of Environment and Forests, Govt. of India, 6624 cases have been filed by the Central Pollution Control Board (CPCB), State Pollution Control Boards (SPCBs) and the Pollution Control Committees (PCCs) of the Union Territories as on 31.10.1997 under the Water and Air Acts. Of these, 2947 cases have been decided and 3677 cases are pending in various courts. The problems in filing the cases are many: a lack of awareness on the part of the public, their fear of the delays in the judicial system and the lack of interest shown by the legal community in taking up environmental cases which earn minimal fees. Obtaining evidence is difficult and no lawyer would risk losing cases on the grounds of lack of evidence, however committed he is to society. The loopholes in environmental laws make it easier for the lawyer to support the cause of the polluter than that of the affected, besides being more remunerative.

All these factors have strengthened the Centre's belief that experience-sharing in this field would be of immense help and inspiration to young lawyers and to various sections of society, particularly NGOs and environmental activists. This publication is a ready guide to the availability and utilisation of existing laws, by affected persons who may not be able to undertake a detailed study of the original. It covers the various heads - Constitutional provisions, the various Acts and some important lists. These information is dealt with in the form of questions and answers to facilitate easy reference.

Over the years, there has been a considerable growth in the field of environmental law. Environmental Tribunals, Green Benches and a National Environmental Appellate Authority have been constituted, while the mandatory Public Hearing is a new and welcome development. Environmental Impact Assessment and Environmental Audit are important developments to control pollution.

India has excellent laws to protect the environment, what is needed is implementation, which is the equal responsibility of the industry and the public. The following pages outline some of the laws and rules which are mandatory before the establishment of an industry.

Key Policies relating to the Environment in India

There are three key policies relating to environmental protection in India. They are:

- The National Forest Policy, 1988
- Policy statement for Abatement of Pollution, 1992
- National Conservation Strategy and Policy Statement on Environment and Development, 1992

'Environment' defined under Indian Law

According to Section 2(a) of the Environmental Protection Act, 1986, 'Environment' includes

- Water, air and land
- The inter-relationship which exists among and between,
- Water, air and land, and
- Human beings, other living creatures, plants, micro-organisms and property

Different Statutes/Legislations enacted in India exclusively for Environmental Protection

The different statutes/legislations enacted in India exclusively for environment protection are

- The Water (Prevention and Control of Pollution) Act, 1974
- The Water (Prevention and Control of Pollution) Rules, 1975
- The Water (Prevention and Control of Pollution) Cess Act, 1977
- The Water (Prevention and Control of Pollution) Cess Rules, 1978
- The Air (Prevention and Control of Pollution) Act, 1981
- The Air (Prevention and Control of Pollution) Rules, 1982
- The Environment Protection Act, 1986
- The Environment (Protection) Rules, 1986
- Hazardous Wastes (Management and Handling) Rules, 1989
- Manufacture, Storage and Import of Hazardous Chemical Rules, 1989
- The Forest Conservation Act, 1980
- The Forest (Conservation) Rules, 1981
- The Wildlife Protection Act, 1972
- The Wild Life (Transactions and Taxidermy) Rules, 1973
- The Wild Life (Stock Declaration) Central Rules, 1973
- The Wild life (Protection) Licensing (Additional Matters for Consideration) Rules, 1983
- The Wild Life (Protection) Rules, 1995

- The Wild Life (Specified Plants - Conditions for Possession by Licensee) Rules, 1995
- The Public Liability Insurance Act, 1991 etc.
- The Public Liability Insurance Rules, 1981
- The National Environment Tribunal Act, 1995
- The National Environment Appellate Authority Act, 1997

What is the difference between the laws enacted before and after Independence with respect to environmental protection in India ?

There are about two hundred laws dealing with environmental protection both before and after independence in India. However, the pre-independence laws have not dealt with environmental protection exclusively. For example, the Indian Penal Code (IPC), 1860, had a chapter (chapter XIV) which dealt with offences affecting public health, safety and convenience, which covered aspects like water, air and noise pollution, whereas the post-independence laws mentioned above deal exclusively with environmental protection.

What are the provisions in the Indian Penal Code for environmental protection?

The Indian Penal Code has a chapter on offences affecting Public Health, Safety, Convenience (Chapter XIV). Sec. 268 provides that "a person is guilty of a public nuisance who does any act or is guilty of an illegal ommission which causes any common injury, danger or annoyance to the public or to the people in general who dwell or occupy property in the vicinity, or which must necessarily cause injury, obstruction, danger, or annoyance to persons who may have occasion to use any public right." The section further explains that a common nuisance is not excusable on the ground that it causes some convenience or advantage. Other concerned provisions are: a "negligent act likely to spread infection or disease dangerous to life" (Sec. 269 I.P.C.), a "malignant act likely to spread infection or disease dangerous to life" (Sec. 270 I.P.C.), "making atmosphere noxious to health" (Sec. 278 I.P.C.).

But the essential requirement of the provision to punish a man is the guilty intention of the accused, i.e. either the act of the accused should be negligent, malignant or voluntary, which vitiates the atmosphere. In case of public nuisance, the Penal Code provides for fines up to Rs.200/- by way of punishment (Sec.290I.P.C.) and for making the atmosphere noxious to health Rs.500/- only (Sec.78 I.P.C.). The punishments are too meagre to meet the objectives. With these penal provisions, it is not possible to check environmental pollution.

Constitution of India

What responsibility does the State / government have towards environmental protection under the Indian Constitution?

The State's responsibility with regard to environmental protection has been laid down under Article 48-A of our Constitution, which reads as follows:

> *"The State shall endeavour to protect and improve the environment and to safeguard the forests and wildlife of the country".*

What responsibility does the citizen have towards environmental protection under our Constitution?

Environmental protection is a fundamental duty of every citizen of this country under Article 51-A(g) of our Constitution which reads as follows:

> *"It shall be the duty of every citizen of India to protect and improve the natural environment including forests, lakes, rivers and wildlife and to have compassion for living creatures."*

What is the importance of Article-21 of the Indian Constitution with respect to environmental protection and we have the right to clean environment ?

Article 21 of the Constitution is a fundamental right which reads as follows:

> *"No person shall be deprived of his life or personal liberty except according to procedure established by law."*

Though this Article does not explicitly mention the environment, the Supreme Court and the various High Courts of the country have given a wider interpretation to the word "life" in this Article. According to the courts, the right to life includes the right to a living environment congenial to human existence.

Why is it that action cannot be taken either against the State or the citizen for violating Articles 48 A and 51 A(g) of the Constitution?

Article 48-A of the Constitution comes under Directive Principles of State Policy and Article 51 A(g) of the Constitution comes under Fundamental Duties.

Unlike Fundamental Rights, violation of Directive Principles of State policy or Fundamental Duties cannot be questioned in a Court of Law.

What responsibility does the State / government have towards raising the level of nutrition and standard of living and improving of public health?

The State's responsibility with regard to raising the level of nutrition and the standard of living and to improve public health has been laid down under Article 47 of the Constitution which reads as follows:

"The State shall regard the raising of the level of nutrition and the standard of living of its people and the improvement of public health as among its primary duties and, in particular, the State shall endeavour to bring about prohibition of the consumption except for medicinal purposes of intoxicating drinks and of drugs which are injurious to health."

What is the importance of the 42nd Amendment to the Constitution?

The 42nd amendment to the Constitution was brought about in the year 1974. Two new Articles were inserted: Art.48-A and Art.51-A(g). The former, under Directive Principles of State Policy, makes it the responsibility of the State Government to protect and improve the environment and to safeguard the forests and wildlife of the country. The latter, under Fundamental Duties, makes it the fundamental duty of every citizen to protect and improve the natural environment including forests, lakes, rivers and wildlife and to have compassion for living creatures.

What are the subjects related to environment in the seventh schedule of the Constitution?

Union List

Entries

52. Industries

53. Regulation and development of oil fields and mineral oil resources

54. Regulation of mines and mineral development

56. Regulation and development of inter-State rivers and river valleys

57. Fishing and fisheries beyond territorial waters.

State List

Entries

6. Public health and sanitation

14. Agriculture, protection against pest and prevention of plant diseases.

18. Land, colonisation, etc.

21. Fisheries

23. Regulation of mines and mineral development subject to the provisions of List-I

24. Industries subject to the provisions of List-I.

Common or Concurrent List

Entries

17A. Forests

17B. Protection of wild animals and birds

20. Economic and social planning

20A. Population control and family planning

As conferred by Article 246(1), while the Union is supreme to make any law over the subjects enumerated in List I, the States, under Article 246 (3), enjoy competence to legislate on the entries contained in List II, and both the Union and the States under Article 246(2) have concurrent jurisdiction on entries contained in List III. In the event of a clash, the Union enjoys a primacy over States in that its legislation in the Union and the Concurrent List prevails over State legislations. Also, the Parliament has residuary powers to legislate on any matter not covered in the three Lists (Art. 248).

Environmental Protection Act, 1986 (E.P.A.)

How are the terms "environment", "environmental pollutant", "environmental pollution" and "hazardous substance" defined under the E.P.A. 1986?

According to Section 2 of *E.P.A.*

a. *"Environment"* includes water, air and land and the inter-relationship which exists among and between water, air and land, and human beings, other living creatures, plants, micro-organism and property.

b. *"Environmental pollutant"* means any solid, liquid or gaseous substance present in such concentration as may be, or tend to be, injurious to environment.

c. *"Environmental pollution"* means the presence in the environment of any environmental pollutant.

d. *"Hazardous substance"* means any substance or preparation which, by reason of its chemical or physics-chemical properties or handling, is liable to cause harm to human beings, other living creatures, plants, micro-organisms, property or the environment.

What are the general powers of the Central government under E.P.A. for the protection and improvement of environment?

Section 3. *Power of Central Government to take measures to protect and improve the environment -*

1. Subject to the provisions of this Act, the Central Government shall have the power to take all such measures as it deems necessary or expedient for the purpose of protecting and improving the quality of the environment pollution.

2. In particular, and without prejudice to the generality of the provisions of sub-section(1), such measures may include measures with respect to all or any of the following matters, namely:

i. Co-ordination of actions by the State Governments, officers and other authorities

a. Under this Act, or the rules made thereunder

or

b. Under any other law for the time being in force which is relatable to the objects of this Act

ii. Planning and execution of a nation-wide programme for the prevention, control and abatement of environmental pollution

iii. Laying down standards for the quality of environment in its various aspects

iv. Laying down standards for emission or discharge of environmental pollutants from various sources whatsoever

Provided that different standards for emission or discharge may be laid down under this clause from different sources having regard to the quality or composition of the emission lr discharge of environmental pollutants from such sources

v. Restriction of areas in which any industries, operations or processes or class of industries, operations or processes shall not be carried out or shall be carried out subject to certain safeguards

vi. Laying down procedures and safeguards for the prevention of accidents which may cause environmental pollution and remedial measures for such accidents

vii. Laying down procedures and safeguards for the handling of hazardous substances

viii. Examination of such manufacturing processes, materials and substances as are likely to cause environmental pollution

ix. Carrying out and sponsoring investigations and research relating to problems of environmental pollution

x. Inspection of any premises, plant, equipment, machinery, manufacturing or other processes, materials or substances and giving, by order, of such directions to such authorities, officers or persons as it may consider necessary to take steps for the prevention, control and abatement of environmental pollution

xi. Establishment or recognition of environmental laboratories and institutes to carry out the functions entrusted to such environmental laboratories and institutes under this Act

xii. Collection and dissemination of information in respect of matters relating to environmental pollution

xiii. Preparation of manuals, codes or guides relating to the prevention, control and abatement of environmental pollution

xiv. Such other matters as the Central Government deems necessary or expedient for the purpose of securing the effective implementation of the provisions of this Act.

3. The Central Government may, if it considers it necessary or expedient so to do for the purposes of this Act, by order published in the Official Gazette, constitute an authority or authorities by such name or names as may be specified in the order for the purpose of exercising and performing such of the powers and functions (including the power to issue directions under Section 5) of the Central Government under this Act and for taking measures with respect to such of the matters referred to in sub-section(2) as may be mentioned in the order and subject to the supervision and control of the Central Government and the provisions of such order, such authority or authorities may exercise the powers or perform the functions or take the measures so mentioned in the order as if such authority or authorities had been empowered by this Act to exercise those powers or perform those functions or take such measures.

Section 4. *Appointment of officers and their powers and functions:*

1. Without prejudice to the provisions of sub-section(3) of Section 3, the Central Government may appoint officers with such designations as it thinks fit for the purpose of this Act and may entrust to them such of the powers and functions under this Act as it may deem fit.
2. The officers appointed under sub-section(1) shall be subject to the general control and direction of the Central Government or, if so directed by that Government, also of the authority or authorities, if any, constituted under sub-section(3) of Section 3 or of any other authority or officer.

Section 5. *Power to give directions:*

Not withstanding anything contained in any other law but subject to the provisions of this Act, the Central Government may, in the exercise of its powers and performance of its functions under this Act, issue directions in writing to any person, officer or any authority and such person, officer or authority shall be bound to comply with such directions.

Explanation : For the avoidance of doubts, it is hereby declared that the power to issue directions under this section includes the power to direct

a. The closure, prohibition or regulation or any industry, operation or process

or

b. Stoppage or regulation of the supply of electricity or water or any other service.

Section 6. *Rules to regulate environmental pollution:*

1. The Central Government may, by notification in the Official Gazette, make rules in respect of all or any of the matters referred to in Section 3.
2. In particular, and without prejudice to the generality of the foregoing power, such rules may provide for all or any of the following matters, namely:

   a. The standards of quality of air, water or soil for various areas and purposes

   b. The maximum allowable limits of concentration of various environmental pollutants (including noise) for different areas

   c. The procedures and safeguards for the handling of hazardous substances

   d. The prohibition and restrictions on the handling of hazardous substances in different areas

   e. The prohibition and restrictions on the location of industries and the carrying on of processes and operations in different areas

   f. The procedures and safeguards for the prevention of accidents which may cause environmental pollution and for providing for remedial measures for such accidents.

Is pollution of land/soil covered under E.P.A.?

Yes. Because under the E.P.A. environment includes water, air and land. The sources of land pollution are:

i. The unintended or incidental pollution of soil with man made chemicals

ii. The spent material from mining, or processing, etc.

iii. The discharge of sewage or waste water from urban areas on the land used for agricultural purposes, particularly that adjoining urban areas

The indiscriminate disposal of solid waste (refuse)

What are the requirements that are to be fulfilled under the E.P.A. by persons carrying on any industry, operation etc.?

According to Section 7, no person carrying on any industry, operation or process shall discharge or emit or permit to be discharged or emitted any environmental pollutant in excess of such standards as may be prescribed.

According to Section 8, no person shall handle or cause to be handled any hazardous substance except in accordance with such procedure and after complying with such safeguards as may be prescribed.

What are the penalties for violations under the E.P.A.?

1. Whoever fails to comply with or contravenes any of the provisions of this Act, or the rules made or orders or directions issued thereunder, shall, in respect of each such failure or contravention, be punishable with imprisonment for a term which may extend to five years or with fine which may extend to one lakh rupees, or with both, and in case the failure or contravention continues, with additional fine which may extend to five thousand rupees for every day during which such failure or contravention continues after the conviction for the first such failure or contravention.
2. If the failure or contravention referred to in sub-section(1) continues beyond a period of one year after the date of conviction, the offender shall be punishable with imprisonment for a term which may extend to seven years.

Can companies and government departments be also prosecuted under E.P.A.?

Yes.

What restriction does the E.P.A. impose on private citizens with respect to courts taking cognizance of offences under the E.P.A. 1986?

Under this Act no court shall take cognizance of any offence except on a complaint made by any person who has given notice of not less than sixty days, in the manner prescribed, of the alleged offence and of his intention to make a complaint, to the Central Government or the authority or officer authorised as aforesaid.

What is the effect of Section 24 of the E.P.A. with respect to other laws that also deal with environmental protection?

According to Section 24, where any act or omission constitutes an offence punishable under this Act and also under any other Act then the offender found guilty of such offence shall be liable to be punished under the other Act and not under this Act.

Air Act

Definition of Air Pollution under the Air Act of 1981

Air pollution means the presence in the atmosphere of any air pollutant.

Air pollutant means any solid, liquid or gaseous substance (including noise) present in the atmosphere in such concentration as may be or tend to be injurious to human beings or other living creatures or plants or property or environment.

Objectives of the Air (Prevention and Control of Pollution) Act 1981

The objective of this Act is to provide for the prevention, control and abatement of air pollution, for the establishment, with a view to carrying out the aforesaid purposes, of Boards, for conferring on and assigning to such Boards powers and functions relating thereto and for matters connected therewith.

Decisions were taken at the United Nations Conference on the Human Environment held in Stockholm in June 1972, in which India participated, to take appropriate steps for the preservation of the natural resources of the earth which, among other things, includes the preservation of the quality of air and control of air pollution.

Therefore it is considered necessary to implement the decisions aforesaid insofar as they relate to the preservation of the quality of air and control of air pollution.

Functions of Central Board under the Air Act

The main functions of the Central Board, as specified in Section 16 of the Act, shall be:

To improve the quality of air and to prevent, control or abate air pollution in the country; and in particular, and without prejudice to the generality of the foregoing functions, the Central Board, may

i. Advise the Central Government on any matter concerning the improvement of the quality of air and the prevention, control or abatement of air pollution
ii. Plan and cause to be executed a nation-wide programme for the prevention, control or abatement of air pollution
iii. Coordinate the activities of the State Boards and resolve disputes among them
iv. Provide technical assistance and guidance to the State Boards, carry out and sponsor investigations and research relating to problems of air pollution and prevention, control or abatement of air pollution
v. Plan and organise the training of persons engaged or to be engaged in programmes for the prevention, control or abatement of air pollution on such terms and conditions as the Central Board may specify
vi. Organise through mass media a comprehensive programme regarding the prevention, control or abatement of air pollution
vii. Collect, compile and publish technical and statistical data relating to air pollution and the measures devised for its effective prevention, control or abatement and prepare manuals, codes, or guides relating to prevention, control or abatement of air pollution
viii. Lay down standards for the quality of air

ix. Collect and disseminate information in respect of matters relating to air pollution

x. Perform such other functions as may be prescribed, under Rules or under an Order.

In addition to the above functions, the Central Board may establish or recognise a laboratory or laboratories to enable the Central Board to perform its functions under this Section efficiently, and it may

a. delegate any of its functions under the Act generally or specially to any of the Committees appointed by it; and

b. do such other things and perform such other acts as it may think necessary for the proper discharge of its functions and generally for the purpose of carrying into effect the purposes of the Act.

Functions of the State Boards under the Air Act 1981

The functions of the State Board, as specified in Section 17, shall be:

a. To plan a comprehensive programme for the prevention, control or abatement of air pollution and to secure the execution thereof

b. To advise the State Government on any matter concerning the prevention, control or abatement of air pollution

c. To collect and disseminate information relating to air pollution

d. To collaborate with the Central Board in organising the training of persons engaged or to be engaged in programmes relating to prevention, control or abatement of air pollution and to organise mass-education programme relating thereto

e. To inspect, at all reasonable times, any control equipment, industrial plant or manufacturing process and to give by order, such directions to such persons as it may consider necessary to take steps for the prevention, control or abatement of air pollution

f. To inspect air pollution control areas at such intervals as it may think necessary, assess the quality of air therein and take steps for the prevention, control or abatement of air pollution in such areas

g. To lay down, in consultation with the Central Board and having regard to the standards for the quality of air laid down by the Central Board, standards for emission of air pollutants into the atmosphere from industrial plants and automobiles or for the discharge of any air pollutant into the atmosphere from any other source whatsoever not being a ship or an aircraft

   Provided that different standards for emission may be laid down under this clause for different industrial plants having regard to the quantity and composition of emission of air pollutants into the atmosphere from such industrial plants

h. To advise the State Government with respect to the suitability of any premises or location for carrying on any industry which is likely to cause air pollution

i. To perform such other functions as may be prescribed or as may, from time to time, be entrusted to it by the Central Board or the State Government

j. To do such other things and to perform such other acts as it may think necessary for the proper discharge of its functions and generally for the purpose of carrying into effect the purposes of the Act.

In addition to the above functions, the State Board may establish or recognise a laboratory or laboratories to enable the State Board to perform its above functions efficiently.

Importance of Sections 19, 20, 21 and 22 of the Air Act

Section 19. *Declaration of air pollution control area:*

The Act has provided for measures which are

a. Preventive in nature, in the case of industries to be established

b. In the case of industries already established they are remedial

The primary responsibility of controlling air pollution is on the Board. The very first measure to be adopted in this respect is the declaration of any area or areas within the State as air pollution control area. The sub-section thus provides that the State Government may, after consultation with the State Board, by notification in the Official Gazette, declare in such manner as may be prescribed, any area or areas within the State as air pollution control area or areas for the purposes of the Act.

As regards power to give instructions for ensuring standards for emission from automobiles, Section 20 of the Act lays down that with a view to ensuring that the standards for emission of air pollutants from automobiles laid down by the State Board under clause(g) of sub-section(1) of Section 17 are complied with the State Government shall, in consultation with the State Board, give such instructions as may be deemed necessary to the concerned authority in charge of registration of motor vehicles under the Motor Vehicles Act, 1988, and such authority shall notwithstanding anything contained in that Act or the rules made thereunder be bound to comply with such instructions.

Penalties for violation of various provisions of the Air Act 1981

Section 37. *Failure to comply with the provisions of section 21 or section 22 or with the directions issued under section 31-A:*

1. Whoever fails to comply with the provisions of section 21 or section 22 or directions issued under section 31-A, shall, in respect of each such failure, be punishable with imprisonment for a term which shall not be less than one year and six months but which may extend to six years and with fine, and in case the failure continues, with an additional fine which may extend to five thousand rupees for every day during which such failure continues after the conviction for the first such failure.
2. If the failure referred to in sub-section(1) continues beyond a period of one year after the date of conviction, the offender shall be punishable with imprisonment with a term which shall not be less than two years but which may extend to seven years and with fine.

Section 38. *Penalties for certain acts:*

1. Whoever Destroys, pulls down, removes, injures or defaces any pillar, post or stake fixed in the ground or any notice or other matter put up, inscribed or placed, by or under the authority of the Board, or

2. Obstructs any person acting under the orders or directions of the Board from exercising his powers and performing his functions under this Act, or
3. Damages any works or property belonging to the Board, or
4. Fails to furnish to the Board or any officer or other employee of the Board any information required by the Board or such officer or other employee for the purpose of this Act, or
5. Fails to intimate the occurrence of the emission of air pollutants into the atmosphere in excess of the standards laid down by the State Board or the apprehension of such occurrence, to the State Board and other prescribed authorities or agencies as required under sub-section(1) of Section 23, or
6. In giving any information which he is required to give under this Act, makes a statement which is false in any material particular, or

For the purpose of obtaining any consent under Section 21, makes a statement which is false in any material particular, shall be punishable with imprisonment for a term which may extend to three months or with fine which may extend to [ten thousand rupees] or with both.

Section 39. *Penalty for contravention of certain provisions of the Act:*

Whoever contravenes any of the provisions of this Act or any order or direction issued thereunder, for which no penalty has been elsewhere provided in this Act, shall be punishable with imprisonment for a term which may extend to three months or with fine which may extend to ten thousand rupees or with both, and in the case of continuing contravention, with an additional fine which may extend to five thousand rupees for every day during which such contravention continues after conviction for the first such contravention.

Companies and government departments be prosecuted under the Air Act

Yes. This is provided under Section 40 and 41.

Section 40. *Offences by companies*

1. Where an offence under this Act has been committed by a company, every person who, at the time the offence was

committed, was directly in charge of, and was responsible to, the company for the conduct of the business of the company, as well as the company, shall be deemed to be guilty of the offence and shall be liable to be proceeded against and punished accordingly: Provided that nothing contained in this sub-section shall render any such person liable to any punishment provided in this Act, if he proves that the offence was committed without his knowledge or that he exercised all due diligence to prevent the commission of such offence.

2. Notwithstanding anything contained in sub-section(1), where an offence under this Act has been committed by a company and it is proved that the offence has been committed with the consent or connivance of, or is attributable to any neglect on the part of any director, manager, secretary or other officer of the company, such director, manager, secretary or other officer shall also be deemed to be guilty of that offence and shall be liable to be proceeded against and punished accordingly.

Explanation:

a. "Company" means any body corporate, and includes a firm or other association of individuals; and
b. "Director", in relation to a firm, means a partner in the firm.

Section 41. *Offences by Government departments:*

1. Where an offence under this act has been committed by any Department of Government, the Head of the Department shall be deemed to be guilty of the offence and shall be liable to be proceeded against and punished accordingly:

   Provided that nothing contained in this section shall render such Head of the Department liable to any punishment if he proves that the offence was committed without his knowledge or that he exercised all due diligence to prevent the commission of such offence.

2. Notwithstanding anything contained in sub-section(1), where an offence under this Act has been committed by a Department of Government and it is proved that the offence has been committed with the consent or connivance of, or is attributable to any neglect on the part of, any officer, other than the Head

of the Department, such officer shall also be deemed to be guilty of that offence and shall be liable to be proceeded against and punished accordingly.

Restriction does the Air Act impose on private citizens with respect to courts taking offence under the Air Act

Under this Act:

No court shall take cognizance of any offence except on a complaint made by any person who has given notice of not less that sixty days, in the manner prescribed, of the alleged offence and of his intention to make a complaint to the Board or officer authorised by the Board.

1. No Court inferior to that of a Metropolitan Magistrate or a Judicial Magistrate of the first class shall try any offence punishable under this Act.
2. Where a complaint has been made by any private citizen the Board shall, on demand by such person, make available the relevant reports in its possession to that person.

The Board may refuse to make any such report available to such person if the same is, in its opinion, against the public interest.

Legal provisions under Indian Law to control automobile pollution

"Automobile" means any vehicle powered either by internal combustion engine or by any method of generating power to drive such vehicle by burning fuel.

Power to give instructions for ensuring standards for emission from automobiles.

As regards power to give instructions for ensuring standards for emission from automobiles, Section 20 of the Act lays down that with a view to ensuring that the standards for emission of air pollutants from automobiles laid down by the State Board under clause(g) of sub-section(1) of Section 17 are complied with, the State Government shall, in consultation with the State Board, give such instructions as may be deemed necessary to the concerned authority incharge of registration of motor vehicles under the Motor Vehicles Act, 1988, and such authority shall notwithstanding anything contained in that Act or the rules made thereunder be bound to comply with such instructions.

Water Act

Definition of Water Pollution under the Water Act 1974

"Pollution" means

1. contamination of water
2. alteration of the physical, chemical or biological properties of water
3. discharge of any sewage or trade effluent or any other liquid, gaseous or solid substance into water (whether directly or indirectly) which may, or is likely to, create a nuisance or render such water harmful or injurious to public health or safety, or to domestic, commercial, industrial, agricultural or other legitimate uses, or to the life and health of animals or plants or of aquatic organisms.

What are the legislations that have been enacted for controlling/ preventing water pollution?

- The Shore Nuisance (Bombay and Kolaba) Act, 1853
- The Orient Gas Company Act, 1857
- Indian Penal Code, 1860
- The Serais Act, 1867
- The North India Canal and Drainage Act, 1873
- The Obstruction in Fairways Act, 1881
- The Indian Easement Act, 1882
- The Indian Fisheries Act, 1897
- The Indian Ports Act, 1908
- The Indian Steam Vessels Act, 1917
- The Poison Act, 1919
- The Indian Forest Act, 1927
- The Damodar valley corporation (Prevention of Pollution of Water) Regulation Act, 1948
- The Factories Act, 1948
- The Mines Act, 1952
- The Orissa River Pollution Act, 1953
- The River Boards Act, 1956

- The Merchant Shipping Act, 1958
- The Maharashtra Prevention of Water Pollution Act, 1969
- The Water (Prevention and Control of Pollution) Act, 1974
- The Water (Prevention and Control of Pollution) Cess Act, 1977.

Objectives of the Water (Prevention and Control of Pollution) Act 1974

The objectives of the Water (Prevention and Control of Pollution) Act are to provide for the Prevention and Control of Water Pollution and the maintenance or restoration of the wholesomeness of water for the establishment, with a view to carrying out the purposes aforesaid, of Boards for the prevention and control of water pollution, for conferring on and assigning to such Boards powers and functions relating thereto and for matters connected therewith.

Functions of the Central (Pollution Control) Board under the Water Act 1974

Functions of Central Board:

1. Subject to the provisions of this Act, the main function of the Central Board shall be to promote cleanliness of streams and wells in different areas of the States.
2. In particular and without prejudice to the generality of the foregoing function, the Central Board may perform all or any of the following functions, namely -
   a. Advise the Central Government on any matter concerning the prevention and control of water pollution
   b. Co-ordinate the activities of the State Boards and resolve disputes among them
   c. Provide technical assistance and guidance to the State Boards, carry out and sponsor investigations and research relating to problems of water pollution and prevention, control or abatement of water pollution
   d. Plan and organise the training of persons engaged or to be engaged in programmes for the prevention, control or abatement of water pollution on such terms and conditions as the Central Board may specify

e. Organise through mass media a comprehensive programme regarding the prevention and control of water pollution [perform such of the functions of any State Board as may be specified in an order made under sub-section (2) of Section 18]

f. Collect, compile and publish technical and statistical data relating to water pollution and the measures devised for its effective prevention and control and prepare manuals, codes or guides relating to treatment and disposal of sewage and trade effluents and disseminate information connected therewith

g. Lay down, modify or annul, in consultation with the State Government concerned, the standards for a stream or well [Provided that different standards may be laid down for the same stream or well or for different streams or wells, having regard to the quality of water flow characteristics of the stream or well and the nature of the use of the water in such stream or well or streams or wells]

h. Plan and cause to be executed a nation-wide programme for the prevention, control or abatement of water pollution

i. Perform such other functions as may be prescribed

3. The Board may establish or recognise a laboratory or laboratories to enable the Board to perform its functions under this section efficiently, including the analysis of samples of water from any stream or well or of samples of any sewage or trade effluents.

Functions of the State Boards under the Water Act 1974

Functions of the State Boards:

1. Subject to the provisions of this Act, the functions of a State Board shall be:
   a. To plan a comprehensive programme for the prevention, control or abatement of pollution of streams and wells in the State and to secure the execution thereof
   b. To advise the State Government on any matter concerning the prevention, control or abatement of water pollution
   c. To collect and disseminate information relating to water pollution and the prevention, control or abatement thereof

d. To encourage, conduct and participate in investigations and research relating to problems of water pollution and prevention, control or abatement of water pollution

e. To collaborate with the Central Board in organising the training of persons engaged or to be engaged in programmes relating to prevention, control or abatement of water pollution and to organise mass education programmes relating thereto

f. To inspect sewage or trade effluents, works and plants for the treatment of sewage and trade effluents and to review plans, specifications or other data relating to plants set up for the treatment of water, works for the purification thereof and the system for the disposal of sewage or trade effluents or in connection with the grant of any consent as required by this Act

g. To lay down, modify or annul effluent standards for the sewage and trade effluents and for the quality of receiving waters(not being water in an inter-State stream) resulting from the discharge of effluents and to classify waters of the State

h. To evolve economical and reliable methods of treatment of sewage and trade effluents, having regard to the peculiar conditions of soils. climate and water resources of different regions and more especially the prevailing flow characteristics of water in streams and wells which render it impossible to attain even the minimum degree of dilution

i. To evolve methods of utilisation of sewage and suitable trade effluents in agriculture) To evolve efficient methods of disposal of sewage and trade effluents on land, as are necessary on account of the predominant conditions of scant stream flows that do not provide for major part of the year the minimum degree of dilution

j. To lay down standards of treatment of sewage and trade effluents to be discharged into any particular stream taking into account the minimum fair weather dilution available in that stream and the tolerance limits of pollution permissible in the water of the stream, after the discharge of such effluents

k. To make, vary or revoke any order

   i. for the prevention, control or abatement of discharges of waste into streams or wells

   ii. requiring any person concerned to construct new systems for the disposal of sewage and trade effluents or to modify, alter or extend any such remedial measures as are necessary to prevent, control or abate water pollution

l. To lay down effluent standards to be complied with by persons while causing discharge of sewage or sullage or both and to lay down, modify or annul effluent standards for the sewage and trade effluents

m. To advise the State Government with respect to the location of any industry the carrying on of which is likely to pollute a stream or well

n. To perform such other functions as may be prescribed or as may, from time to time, be entrusted to it by the Central Board or the State Government.

2. The Board may establish or recognise a laboratory or laboratories to enable the Board to perform its functions under this section efficiently, including the analysis of samples of water from any stream or well or of samples of any sewage or trade effluents.

Importance of Section 24 of the Water Act 1974

According to Section 24 of the Water Act, 1974

a. No person should knowingly cause or permit any poisonous, noxious or polluting matter determined in accordance with such standards as may be laid down by the State Board to enter (whether directly or indirectly) into any stream or well or sewer or on land; or

b. No person shall knowingly cause or permit to enter into any stream any other matter which may tend, either directly or in combination with similar matters, to impede the proper flow of the water of the stream in a manner leading or likely to lead to a substantial aggravation of pollution due to other causes or of its consequences.

However, a person shall not be guilty of an offence under sub-section (1. by reason only of having done or caused to be done any of the following acts, namely

a. Constructing, improving or maintaining in or across or on the bank or bed of any stream, any building, bridge, weir, dam, sluice, dock, pier, drain or sewer or other permanent works which he has a right to construct, improve or maintain
b. Depositing any materials on the bank or in the bed of any stream for the purpose of reclaiming land or for supporting, repairing or protecting the bank or bed of such stream provided such materials are not capable of polluting such stream
c. Putting into any stream any sand or gravel or other natural deposit which has flowed from or been deposited by the current of such stream
d. Causing or permitting, with the consent of the State Board, the deposit accumulated in a well, pond or reservoir to enter into any stream.

Penalty for contravention of provisions of Section 24:

Whoever contravenes the provisions of Section 24 shall be punishable with imprisonment for a term which shall not be less than one year and six months but which may extend to six years and with fine.

Municipal Corporation, Companies, Government departments also be prosecuted under the Water Act

Yes. This is provided under sections 47 and 48 of the Water Act. Section 47. *Offences by companies:*

1. Where an offence under this Act has been committed by a company every person who at the time the offence was committed was in charge of, and was responsible to the company for the conduct, of the business of the company, as well as the company, shall be deemed to be guilty of the offence and shall be liable to be proceeded against and punished accordingly:

Provided that nothing contained in this sub-section shall render any such person liable to any punishment provided in this Act if he proves that the offence was committed without his knowledge or that he exercised all due diligence to prevent the commission of such offence.

2. Notwithstanding anything contained in sub-section(1., where an offence under this Act has been committed by a company and it is proved that the offence has been committed with the consent or connivance of, or is attributable to any neglect on the part of, any director, manager, secretary or other officer shall also be deemed to be guilty of that offence and shall be liable to be proceeded against and punished accordingly.

   Explanation : For the purposes of this section

   a. "Company" means any body corporate, and includes a firm or other association of individuals

   b. "Director" in relation to a firm means a partner in the firm.

Section 48. *Offences by Government Departments.*

Where an offence under this Act has been committed by any Department of Government, the head of the Department shall be deemed to be guilty of the offence and shall be liable to be proceeded against and punished accordingly.

Emergency measures can the Central / State pollution boards take under the Water Act

Emergency measures in case of pollution of stream or well:

1. Where it appears to the State Board that any poisonous, noxious or polluting matter is present in [any stream or well or on land by reason of the discharge of such matter in such stream or well or on such land] or has entered into that stream or well due to any accident or other unforeseen act or event, and if the Board is of opinion that it is necessary or expedient to take immediate action, it may for reasons to be recorded in writing, carry out such operations as it may consider necessary for all or any of the following purposes, that is to say,

a. Removing that matter from the [stream or well or on land] and disposing it off in such manner as the Board considers appropriate
b. Remedying or mitigating any pollution caused by its presence in the stream or well
c. Issuing orders immediately restraining or prohibiting the person concerned from discharging any poisonous, noxious or polluting matter [into the stream or well or on land], or from making in sanitary use of the stream or well.

2. The power conferred by sub-section(1) does not include the power to construct any works other than works of a temporary character which are removed on or before the completion of the operations.

What are the different powers given to the Central/ State Boards under the Water Act?

The powers given to Central/State Boards to make application to courts for restraining apprehended pollution of water in streams or wells:

1. Where it is apprehended by a Board that the water in any stream or well is likely to be polluted by reason of the disposal or likely disposal of any matter in such stream or well or in any sewer or on any land, or otherwise, the Board may make an application to a court, not inferior to that of a Metropolitan Magistrate or a Judicial Magistrate of the first class, for restraining the person who is likely to cause such pollution from so causing
2. On receipt of an application under sub-section(1. the court may make such order as it deems fit.
3. Where under sub-section(2) the court makes an order restraining any person from polluting the water in any stream or well, it may in that order
   i. Direct the person who is likely to cause or has caused the pollution of the water in the stream or well, to desist from taking such action as is likely to cause pollution or, as the case may be, to remove from such stream or well, such matter

ii. Authorise the Board, if the direction under Clause(i. (being a direction for the removal of any matter from such stream or well. is not complied with by the person to whom such direction is issued, to undertake the removal and disposal of the matter in such manner as may be specified by the court.

4. All expenses incurred by the Board in removing any matter in pursuance of the authorisation under clause(ii. of sub-section(3. or in the disposal of any such matter may be defrayed out of any money obtained by the Board from such disposal and any balance outstanding shall be recoverable from the person concerned as arrears of land revenue or of public demand.

Section 33-A. *Power to give directions:*

Notwithstanding anything contained in any other law, but subject to the provisions of this Act and to any directions that the Central Government may give in this behalf, a Board may, in the exercise of its powers and performance of its functions under this Act, issue any directions in writing to any person, officer or authority, and such person, officer or authority shall be bound to comply with such directions.

Explanation: For the avoidance of doubts, it is hereby declared that the power to issue directions under this section includes the power to direct

a. Closure, prohibition or regulation of any industry, operation or process

   or

b. The stoppage or regulation of supply of electricity, water or any other service.

Restriction under the Water Act impose on private citizens with respect to courts taking cognizance of offences under the Water Act

Under this act

1. No court shall take cognizance of any offence except on a complaint made by any person who has given notice of not less than sixty days, in the manner prescribed, of the alleged

offence and of his intention to make a complaint to the Board or officer authorised by the Board.

N.B.

1. No court inferior to that of a Metropolitan Magistrate or a Judicial Magistrate of the first class shall try any offence punishable under this Act
2. Where a complaint has been made by any private citizen the Board shall, on demand by such person make available the relevant reports in its possession to that person.

The Board may refuse to make any such report available to such person if the same is, in its opinion, against the public interest.

Provisions to safeguard rivers and (natural lakes)

According to (Ministry of Environment and Forests, Government of Tamil Nadu) GO No 1 dated 6-2-84, of the a factory can come up at a distance of one kilometre from a river or natural water body. The Ministry of Environment and Forests passed a GO No. 213 dated 30-3-89 amending the distance from one kilometre to five kilometres.

The Ministry of Environment and Forest, Government of Tamil Nadu, in its GO No 127 dated 8-5-98, has amended its earlier reference with respect the river. (GO No 213, Ministry of Environment and Forest, dated 30-8-89. that no industry coming under red category should be located within 5 km from major river Cauvery and its tributaries.

Forest Act

What are the requirements that should be met before declaring an area as a Wildlife Sanctuary or a National Park under the Forest Acts?

Section 29. *Protected forests:*

1. The State Government may, by notification in the Official Gazette declare the provisions of this chapter applicable to any forest-land or waste-land which is not included in a reserved forest, but which is the property of the Government, or over which the Government has proprietary rights, or to the whole or any part of the forest produce of which the Government is entitled.

2. The forest-land and waste-land comprised in any such notification shall be called a "protected forest".
3. No such notification shall be made unless the nature and extent of the rights of Government and of private persons in or over the forest-land or waste-land comprised therein have been inquired into and recorded at a survey or settlement, or in such other manner as the [State Government] thinks sufficient. Every such record shall be presumed to be correct until the contrary is proved:

   Provided that if, in the case of any forest-land or waste-land, the [State Government] thinks that such enquiry and record are necessary but that they will occupy such length of time as in the meantime to endanger the rights of Government, the [State Government] may, pending such inquiry and record, declare such land to be a protected forest, but so as not to abridge or affect rights of individuals or communities. Of the control over Forests and Lands not being the Property of Government.

Section 35. *Protection of forests for special purposes:*

1. The State Government may, by notification in the official Gazette, regulate or prohibit in any forest or waste-land:
   a. The breaking up or clearing of land for cultivation
   b. The pasturing of cattle
   c. The firing or clearing of the vegetation

   When such regulation or prohibition appears necessary for any of the following purposes:
   i. For protection against storms, winds, rolling stones; floods and avalanches
   ii. For the preservation of the soil on the ridges and slopes and in the valleys or hilly tracts, the prevention of landslips or of the formation of ravines and torrents, or the protection of land against erosion or the deposit thereon of sand, stones or gravel
   iii. For the maintenance of a water supply in springs, rivers and tanks
   iv. For the protection of roads, bridges, railways and other lines of communication

v. For the preservation of public health.

2. The State Government may, for any, such purpose, construct as its own expense, in or upon any forest or waste-land, such work as it thinks fit.

3. No notification shall be made under sub-section(1. nor shall any work be begun under sub-section(2. until after the issue of a notice to the owner of such forest or land calling on him to show cause, within a reasonable period to be specified in such notice, why such notification shall not be made or work constructed, as the case may be and until his objections, if any, and any evidence he may produce in support of the same, have been heard by an officer duly appointed for that purpose and have been considered by the State Government.

Changes in the Forest Conservation Act, 1980

Breaking up of the soil or the clearing of the forest land seriously affects reafforestation or regeneration of forest. Therefore, such breaking up of the soil can only be permitted after taking into consideration all aspects of the question, such as the overall advantages and disadvantages to the economy of the country, environmental conditions, the ecological imbalances, that are likely to occur, the effects on the flora and the fauna in the area, etc. It was, therefore, thought that the entire control of forest areas should vest in the Central Government. With that end in view, Sec.2 provided that prior approval of the Central Government should be obtained before permitting the use of the forest land for non-forest purposes.

Meaning of "non-forest purpose" under the Forest Conservation Act, 1980

"Non-forest purpose" means the breaking up or clearing of any forest land or portion thereof for

a. The cultivation of tea, coffee, spices, rubber, palms, oil- bearing plants, horticultural crops or medicinal plants;

b. Any purpose other than re-afforestation, but does not include any work relating or ancillary to conservation, development and management of forest and wild life, namely, the establishment of check-posts, firelines, wireless communications and construction of fencing, bridges and culverts, dams, water-holes, trench marks, boundary marks, pipelines or other like purposes.

Wildlife Protection Act

Definition of "Wildlife" under the Wildlife Protection Act, 1972

According to the Wildlife Protection Act, 1972 "wildlife" includes any animal, bees, butterflies, crustacea, fish and moths; and aquatic or land vegetation which forms part of any habitat.

Protection of Wildlife

The Wildlife Protection Act was passed by the Indian Parliament in the year 1972 to protect India's wildlife. However, in the 20 years that have passed since the Act came into force, the number of wild animals is going down alarmingly, despite of Government efforts to protect them. With the increase in population, there is greater pressure on land. Forests are being destroyed as human habitations expand, thereby shrinking the habitats of our wildlife. There is also the clandestine international trade in wildlife and wildlife products which is a major cause for their wanton destruction. Meanwhile, the growing consumer society and the increasing emphasis on luxury and vanity items has also caused the exploitation of wildlife in the name of industrial progress.

The major task of protecting wildlife cannot be handled by the Government machinery alone through its limited officials, but should be the duty of every individual. This was one of the reasons why a new provision, Article 51 A (g., was inserted into our Constitution, making it the fundamental duty of every citizen to protect and improve the natural environment, including forests, lakes, rivers and wildlife, and to have compassion for living creatures.

Points To Remember While Making A Complaint Under The Wildlife (Protection Act)

- Of foremost importance is the correct and full identification of the "wildlife" species involved in the offence, which decides all further action.

- Provisions of the Act do not apply if the species involved does not occur wild in India (except ivory of the African elephant) or is of farm origin.
- Keep with you always, for ready reference, an up-to-date copy of the Wildlife (Protection. Act and Wildlife (Protection. (Tamil Nadu) Rules. (The Police effecting the arrest/seizure at your instance, as often also the Magistrate (who takes note of the arrest/seizure as per Section 50(5.), might wish to refer to the Act for guidance.
- If you are a Hon. Wild llife Warden, you would be well advised to always keep with you your identity card (as Hon. Wild life Warden.
- If possible, you may detain an offender against The Wildlife Act caught in the act, but hand him over to the Police with the least possible delay (Section 43 of the Criminal Procedure Code.
- Specify the nature of the offence in your complaint so that the same is recorded by the police in the First Information Report (FIR).
- As you wait for the police or the Wildlife Officer, please ensure that evidence of the offence is guarded, to be gathered/recorded by the authorised official.
- It would help effective prosecution if the NGO representative becomes a party to the Panchanama/mahazar1 so that in the court trial against the accused, he could be useful as a witness.

Remember

i. Fix priorities in needed action.

ii. Keep in mind the many constraints of the very few wildlife officials to enforce the law. Help them concentrate on priority areas by not diverting their attention to less important offences.

Important points which must be incorporated in a Complaint to the Police / Forest officer as also in a Panchanama / Mahazar:

Whereas the Wildlife or Police Officer dealing with the case would almost always get this statement recorded in a manner acceptable to the courts, it would be good for the NGO to know the salient features of the documents so as to ensure a better recording of the evidence.

a. Name(s),occupation and full postal addresses of the witnesses(should not be less than 2 witnesses in any case.
b. The place, date, time of commencement and completion of the document.
c. The full names and other details of the other persons present at scene/detection of crime including those detained / arrested.
d. The exact species with the Latin name, preferably, identifying the species as the same as that figuring in a particular schedule of the Wildlife Act.
e. A detailed and accurate description of the animal / trophy / animal article together with any identification mark that you may affix on it to help later recognition. (Cases have been lost in the court due to failure on this count)
f. Please ensure that evidence of the offence is gathered and recorded in a manner that will stand judicial scrutiny during the trial.

Panchanama/Mahazar : This is an important document in any prosecution as much as it is a record of the facts as observed by the witnesses.

Coastal Regulation Zone Notification

Law / Regulation to Protect the Coast

Yes. A notification under Section 3(1. and section 3(2.(v) of the Environment (Protection. Act, 1986 and the Rule 5(3. (d. of the Environment (Protection. Rules, 1986 declaring the coastal stretches as Coastal Regulation Zone (CRZ) and imposing restrictions on industries, operations and processes in the CRZ was published vide S.O.No. 944(E., dated 15th December, 1990.

What are the prohibited activities under CRZ Notification?

The following activities are declared as prohibited within the Coastal Regulation Zone:

a. Setting up new industries and expansion of existing industries except those directly related to waterfront or directly needing foreshore facilities.

b. Manufacture or handling or storage or disposal of hazardous substances as specified in the Notifications of the Government of India in the Ministry of Environment & Forests. No.S.O.594(E. dated 28th July, 1989, S.O.966(E. dated 27th November, 1989 and GSR 1037(E. dated 5th December 1989,2. except transfer of hazardous substances from ships to ports, terminals and refineries and vice versa in the Port areas "Provided that the Government of India in the Ministry of Surface Transport, on a case to case basis. Appended to this notification within the existing port limits of existing port limits of existing ports and harbours and in those areas of ports that have not been classified as CRZ-I subject to implementation of safety regulations including guidelines issued by Oil Safety Directorate in the Government of India, Ministry of Petroleum and Natural Gas after ensuring proper location of site and availability of necessary equipment to meet the safety norms and exigencies arising due to any accident or spillage.

c. Setting up and expansion of fish processing units (including warehousing (excluding hatchery and natural fish drying in permitted area.

d. Setting up and expansion of units/mechanisms for disposal of waste and effluents, except facilities required for discharging treated effluents, into the water course with approval under the Water (Prevention and Control of Pollution. Act, 1974, and except for storm water drains

e. Discharge of untreated wastes and effluents from industries, cities or towns and other human settlements. Schemes shall be implemented by the concerned authorities for phasing out the existing practices, if any, within a reasonable time period not exceeding three years from the date of this Notification

f. Dumping of city or town waste for the purpose of land-filing or otherwise; the existing practice, if any, shall be phased out within a reasonable time not exceeding three years from the date of the Notification

g. Dumping of ash or any wastes from thermal power stations

h. Land reclamation, bundling or disturbing the natural course of sea water, except those required for construction of ports, jetties, wharves, quays, slip ways, bridges and sea-links and for other facilities that are essential for activities permissible under the notification or for control of coastal erosion and maintenance or cleaning of waterways, channels and ports and for prevention of sandbars for tidal regulators, storm water drains and structures or for prevention of salinity ingress and for sweet water recharge

i. Mining of sands, rocks and other substrata materials, except those rare minerals not available outside the CRZ areas

j. Harvesting of drawal of ground water and construction of mechanisms therefore within 200m of HTI [Provided that drawal of ground water where no other source of water is available and when done manually through ordinary wells or hands pumps, for drinking and domestic purposes, in zone between 50 to 200m or the CRZ, whichever is less from High Tide Line in case of rivers, creeks and backwaters subject to such restrictions, as may be deemed necessary, in areas affected by sea water intrusion, that may be imposed by an authority designated by State Government/Union Territory/Administration.]

k. Construction activities in ecologically sensitive areas of this Notification

l. Any construction activity between the Low Tide Line and High Tide Line except facilities for carrying treated effluents and waste water discharges into the sea, facilities for carrying sea water for cooling purposes, oil, gas and similar pipelines and facilities essential for activities permitted under the notification

m. Dressing or altering of sand dunes, hills, natural features including landscape changes for beautification, recreational and other such purpose, except as permissible under this Notification

n. Nothing contained in this paragraph shall apply to aqua culture.

Permissible Activities Under the CRZ Notification and their Regulation

All other activities, except those mentioned in the question above, will be regulated as under:

1. Clearance shall be given for any activity within the Coastal Regulation Zone only if it requires waterfront and foreshore facilities;
2. The following activities will require environmental clearance from the Ministry of Environment & Forests, Government of India, namely;
   i. Construction activities related to Defence requirements for each which foreshore facilities are essential (e.g. slipways, jetties, etc.); except for classified operational component of defence projects for which a separate procedure shall be followed. (Residential buildings, office buildings, hospital complexes, workshops shall not come within the definition of operational requirements except in very special cases and hence shall not normally be permitted in the CRZ).
   ii. Operational constructions for ports and harbours and light houses construction for activities such as jetties, wharves, quays, and slipways :

      Provided that for expansion or modernisation of existing ports and harbours including fishing harbours operational constructions for ports and harbours and construction of jetties, wharves, quays, slipways. Single Point Mooring and Single Buoy Mooring and for reclamation for facilities essential for operational requirements of ports and harbours in areas with in the existing port limits, except the areas classified as category CRZ-I (1., shall require environmental clearance from Government of India in the Ministry of surface Transport, which shall take decision on these activities on the basis of Environmental impact assessment Report.

      Provided further that reclamation for commercial purposes such as shopping and housing complexes, hotels and entertainment activities shall not permissible.
   iii. Thermal power plants (only foreshore facilities for transport of raw materials facilities for intake of cooling water); and out fall for discharge of treated waste water cooling water.
   iv. All other activities with investment exceeding rupees five crores.

National Environmental Appellate Authority, Environmental Tribunals and Green Benches

National Environmental Appellate Authority Act

Purpose with which the National Environmental Appellate Authority Act 1997 has been enacted: The National Environmental Appellate Authority Act has been brought with the purpose of providing for the establishment of a National Environmental Appellate Authority to hear appeals with respect to restriction of areas in which any industries, operations or processes or class of industries, operations or processes shall not be carried out or shall be carried out subject to certain safeguards under the Environment (Protection. Act, 1986 and for matters connected therewith or incidental thereto.

Uniqueness of National Environmental Appellate Authority in dealing with appeals before them: According to Section 12 of the Act the Authority shall not be bound by the procedure laid down in the code of civil procedure, 1908, but shall be guided by the principles of natural justice. Subject to the other provisions of this Act and of any rules made by Central Government, the Authority shall have power to regulate its own procedure, including the fixing of places and times of its enquiry and deciding whether to sit in public or private. Also, with the effect from the date of establishment of the Authority, no civil court or other authority shall have jurisdiction to entertain any appeal in respect of any matter with which the Authority is empowered by or under this Act.

National Environmental Tribunals Act

Objectives with which the National Environmental Tribunals Act, 1995 has been enacted: The National Environmental Tribunal Act has been enacted to provide for strict liability for damages arising out of any accident occurring while handling any hazardous substance and for the establishment of a National Environment Tribunal for effective and expenditures disposal of cases arising from such accidents, with a view to giving relief and compensation for damages to persons, property and the environment and for matters connected therewith or incidental thereto.

Green Benches

Definition of Green benches: Green benches are those constituted by the Chief Justice of the respective High Courts either on their own or on directions from the Chief Justice of the Supreme Court to constitute exclusively a bench (quorum consisting of more than one Judge. to deal with matters relating to environment and connected there with. The Green Bench in the respective High Courts deals with matters relating to Environment either on a particular day of the week exclusively or when and where the situation demands immediate action. West Bengal and Tamil Nadu are examples of some states which have constituted Green Benches.

Environmental Impact Assessment

EIA (Environmental Impact Assessment), and Its legal sanctity

Environmental Impact Assessment (EIA. is an important management tool for ensuring optimal use of natural resources for sustainable development, and was introduced in India initially for River Valley Projects in 1978-79. The scope of the EIA has been enhanced to cover other developmental sectors such as industries, mining schemes, energy, etc. To facilitate project proponents in collection of environmental data and formulation of environmental management plans, it is now mandatory under the Environment (Protection. Act, 1986, for 29 categories of developmental activities involving investment beyond certain thresholds.

The notification was issued on 27th January 1994 and was amended on 4th May 1994. This, it is hoped would provide an opportunity both for the project proponents and Government to assess the impact of the concerned project on the environment before it actually comes into play.

Purpose of Environmental Impact Assessment

The purpose of Environmental Impact Assessment (EIA). is to identify and evaluate the potential impacts (beneficial and adverse. of development projects on the environmental system. It is an useful aid for decision making based on understanding of the environmental implications including social, cultural and aesthetic concerns which could be integrated with the analysis of the project costs and benefits. This exercise should be undertaken early enough at the planning stage of projects for selection of environmentally compatible sites, process technologies and such other environmental safeguards.

While all industrial projects may have some environmental impacts all of them may not be significant enough to warrant elaborate assessment procedures. The need for such exercises will have to be decided after initial evaluation of the possible implications of a particular project and its location. The projects which could be the candidates for detailed Environmental Impact Assessment include:

i. Those which can significantly alter the landscape, land use pattern and lead to concentration of working and service population

ii. Those which need upstream development activity like assured mineral and forest products supply or downstream industrial process development

iii. Those involving manufacture, handling and use of hazardous materials

iv. Those which are sited near ecologically sensitive areas, urban centres, hill resorts, places of scientific and religious importance

v. Industrial estates with constituent units of various types which could cumulatively cause significant environmental damage.

Environmental Statement

Definition of Environmental Statement: It is defined by the International Chamber of Commerce as "a management tool comprising a systematic, documented, periodic and objective evaluation of how well environmental organisations, management and equipment are performing with the aim of helping to safeguard the environment by:

a. Facilitating management control of environmental protection

b. Assessing compliance with company policies which, would include muting regulatory requirements.

Industries which require Environment Auditing

A gazette notification on environmental audit had been issued by the Ministry of Environment and Forests on 13th March, 1992 (amended vide notification GSR 386 (E. dated 22 April, 1993.. This notification applies to every person carrying on an industry, operation or process requiring consent to operate under Section 25 of the Water (Prevention and Control of Pollution. Act, 1974 (6 of 1974. or under section 21 of the Air (Prevention and Control of Pollution. Act, 1981 (14 of 1981., or both, or authorization under the Hazardous Wastes (Management and Handling. Rules, 1989, issued under the Environment (Protection. Act, 1986 (29 of 1986.. The notification requires that an Environmental Statement for the financial year ending on the 31st March be submitted to the concerned State Pollution Control Board on or before the 30th September of the same year.

Public Hearings

Definition of Public Hearings

Public hearings are those hearings which are required under the law before a sanction is given by the Government in respect of any project which falls under the 29 categories of activities (Annexure-3. which require environmental clearance from the Ministry of Environment and Forests, Government of India (MoEF)

Public Hearings are a mandatory requirement under the law, and provide an opportunity for the public to get to know about the coming up of any new project falling under the 29 categories of activities requiring environmental clearance of the MoEF,GOI, and also an opportunity where the concerns, suggestions, views, comments and objections of the public are heard by the public hearing panel.

Legal sanctity of a Public Hearing

The EIA (Environmental Impact Assessment) Notification of January 27,1994 has been amended on April 10, 1997. By this amendment, a Public Hearing is mandatory for all the 29 categories of activities which require environmental clearance from the MoEF.

Various Stages involved in Conducting a Public Hearing

Stage-1: The State Pollution Control Board (SPCB. shall cause a notice of Environmental Public Hearing by publishing the same in at least two newspapers widely circulated in the region around the project, one of which shall be in the vernacular language of the locality concerned.

Stage-2: The State Pollution Control Board (SPCB. shall mention the date, time and the venue of the public hearing and also the name and address of the industry/unit proposed to come up for which a clearance is sought.

Stage-3: From the date of publication of the Notification, 30 days time is provided to the public inviting their suggestions, views, comments and objections to the said project.

Stage-4: The public are provided access to the executive summary of the Environmental Impact Assessment project during this one month period at the following places:

1. District Collector's office
2. District Industry Centre
3. Office of the Chief Executive Officers of the Zilla Parishad or Commissioner of the Municipal Corporation / Local body as the case may be
4. Head office of concerned State Pollution Control Board and its concerned Regional Office
5. Department of State Government dealing with the subject of Environment.

Stage-5: The actual public hearing is conducted as per the date, time and venue mentioned in the notification where the concerns of the public are heard by the Public Hearing Panel constituted for the purpose.

Can any member of the public participate in a Public Hearing?

Though the provision mentioned under sub-paragraph [ii] of paragraph 2 of Schedule 4 of EIA Notification says that all persons, including bonafide residents, environmental groups and others located at the project site / sites of displacement / sites likely to be affected can participate in the public hearing, the explanation provided in the same paragraph explains the word "person" as

a. Any person who is likely to be affected by the grant of environmental clearance
b. Any person who owns or has control over the project with respect to which an application has been submitted for environmental clearance
c. Any association of persons whether incorporated or not likely to be affected by the project and/or functioning in the field of environment
d. Any local authority within any part of whose local limits is within the neighbourhood wherein the project is proposed to be located.

Therefore, because of the explanation, only those who fall within the categories (a) to (d) can participate in a public hearing

Members of a Public Hearing Panel

The members of a public hearing panel are:

a. Representative of State Pollution Control Board
b. District Collector or his nominee
c. Representative of State Government dealing with the subject of Power
d. Representative of Department of State Government dealing with Environment
e. Not more than three representatives of the local bodies such as Municipalities or Panchayats
f. Not more than three senior citizens of the area nominated by the District Collector.

ECOMARK

Definition of ECOMARK: The ECOMARK is a label given to an Environment Friendly Product. Household and other consumer products which meet certain environmental criteria along with the quality requirements of the Indian Standards Institute for that product may be accredited and labeled under this scheme.

Environment Friendly Product

Any product which is made, used or disposed off in a way that significantly reduces the harm it would otherwise cause the environment could be considered an Environment Friendly Product.

Issues in ECOMARK notifications

The Ministry of Environment and Forests, Government of India, issues the ECOMARK notifications.

Number of notifications issued till date

Till date, 18 notifications have been issued by the Ministry of Environment and Forests on different products criteria.

Other Environment - Related Laws

The Shore Nuisance (Bombay and Kolaba. Act, 1853

This is the earliest Act on the statue book concerning control of water pollution in India.

The Serais Act, 1867

The Act enjoined upon a keeper of Serai or an inn to keep a certain quality of water fit for consumption by "persons and animals using it" to the satisfaction of the District magistrate or his nominees. Failure for maintaining the standard entailed a liability of rupees twenty.

The North India Canal and Drainage Act, 1873

Certain offences have been listed under the Act contained in Section 70.

Obstruction in Fairways Act, 1881

Section 8 of the Act empowered the Central Government to make Rules to regulate or prohibit the throwing of rubbish in any fairway leading to a port causing or likely to give rise to a bank or shoal.

Indian Easements Act, 1882

Illustrations (f), (h) and (j) of Section 7 of the Act deal with pollution of waters.

The Indian Fisheries Act, 1897

The Indian Fisheries Act, 1897 contains seven sections. Section 5 of the Act prohibits destruction of fish by poisoning waters.

Indian Ports Act, 1908

Water pollution by oil has been regulated by the Indian Ports Act, 1908.

The Indian Forest Act, 1927

Section 26(i. of the Act makes it punishable if any person, who, in contravention of the rules made by the State Government, poisons water of a forest area. The State Government has been empowered under Section 32(f. to make rules relating to poisoning of water in forests.

The Damodar Valley Corporation Act, 1948

The Act authorises the Corporation to make regulations with the previous sanction of the Central Government for preventing "pollution of water".

The Factories Act, 1948

Factories Act, 1948 is a social welfare legislation intend to secure health, safety and welfare of the workers employed in factories. However, some of the provisions of this Act are concerned with prevention of water pollution.

The Mines Act, 1952

Chapter V of the Act deals with provisions regarding health and Safety of the employees. Section 19(i. emphasises upon arrangement for the quality of water for drinking purposes.

The River Boards Act, 1956

The Act provides for the creation of River Boards for regulation and development of interstate rivers and river valleys. One of the functions of the Board is to advise to the government concerned on "prevention of pollution of the waters of the interstate rivers".

The Merchant Shipping Act, 1958

The International Convention for the Prevention of Pollution of the Sea by Oil, 1954 is the first treaty for the reduction of oil pollution of the sea. In order to give effect to this Convention, the Merchant Shipping Act regulates and controls the discharge of oil or oil mixture by an Indian tanker or ship within any of the prohibited zones or by a foreign tanker or other ship within the prohibited zone adjoining the territories of India. Further, there is a prohibition for discharging any oil anywhere at sea from an Indian ship.

Some Important Cases on Environment

RATLAM CASE

Municipal Council, Ratlam v. Vardhichand,
AIR 1980 SC 1622

DEHRADUN QUARRYING CASE

Rural Litigation and Entitlement Kendra v. State of UP.,
AIR 1987 SC 2187

SHRI RAM GAS LEAK CASE

M.C. Mehta v. Union of India, AIR 1987 SC 965
M.C. Mehta v. Union of India, AIR 1987 SC 982
M.C. Mehta v. Union of India, AIR 1987 SC 1086

RADIATION CASE OR IRISH BUTTER CASE

Dr. Shiva Rao Shantaram Wagle v. Union of India, AIR 1988 SC 952

GANGA POLLUTION (TANNERIES) CASE

M.C. Mehta v. Union of india, AIR 1988 SC 1037

GANGA POLLUTION (MUNICIPALITIES) CASE

M.C. Mehta v. Union of India, AIR 1988 SC 11 15

BHOPAL CASE

Union Carbide Corporation v. Union of India, AIR 1990 SC 273
Charan Lal Sah V. Union of india, AIR 1990 SC 1480

TAJ MAHAL CASE

M. C. Mehta v. Union of India & Others

Writ petition (civil no) 13381 of 1984 Supreme Court decided on December 30,1996

(Kuldip Singh, J. and Faizaluddin, J.

Forests

T.N.Godavarman Tirumulpad Vs Union of India and others: AIR 1997 SC 1233.

Aqua Culture

S. Jagannath Vs Union of India (1997) 2 SCC 87: AIR 1997 SC 811.

Tannery Pollution

Vellore Citizens Welf are Forum Vs Union of India (1996) 5 SCC 647.

Indiscriminate withdrawal of underground water

M.C. Mehta Vs Union of India, (1997) 11 SCC 312.

Shifting/Relocation/Closure of Hazarduous/Noxious/Heavy/ Large Industries from Delhi

M.C. Mehta Vs Union of India, (1997) 11 SCC 327.

Public Trust Doctrine

M.C. Mehta Vs Kamal Nath, (1997) 1 SCC 388.

Precautionary Principle and Principle of Sustainable Development

M.C. Mehta (Badkhal and Surajkund Lakes Matter) Vs Union of India (1997) 3 SCC 715.

Annexure - 1

Acts for the Protection of the Indian Environment Since 1857

- The Orient Gas Company Act, 1857
- The Serais Act, 1867
- The Northern India Canal and Drainage Act, 1873
- The Obstruction in Airways Act, 1881
- The Indian Fisheries Act, 1897
- The Indian Ports Act, 1901
- The Bengal Smoke Nuisance Act, 1905
- The Explosives Act, 1908
- The Bombay Smoke Nuisance Act, 1912
- The Inland Stream Vessel Act, 1917
- The Mysore Destructive Insects & Pests Act, 1917
- The Poison Act, 1919
- The Andhra Pradesh Agricultural, Pest & Diseases Act, 1919
- The Indian Boilers Act, 1923
- The Workmen's Compensation Act, 1923
- The Indian Forest Act, 1927
- The Motor Vehicles Act, 1939
- The Bihar Wastelands (Reclamation, Cultivation & Improvement) Act, 1946
- The Mines and Minerals (Regulation and Development) Act 1947
- The Damodar Valley Corporation (Prevention of Pollution of Water) Regulation Act, 1948
- The Factories (Pollution and Pesticides) Act, 1948
- The Employees State Insurance Act, 1948
- The Andhra Pradesh Improvement Schemes (Land utilisation. Act, 1949
- The Industries (Development and Regulation. Act, 1951
- The Calcutta Municipal Act, 1951
- The Madhya Pradesh Control of Music and Noises Act, 1951

- The Maharashtra Prevention of Water Pollution Act, 1953
- The Shore Nuisance (Bombay and Colaba. Act, 1953
- The Orissa River Pollution & Prevention Act, 1953
- The Assam Agricultural Pests and Disease Act, 1954
- The Prevention of Food Adultertion Act, 1954
- The U.P. Agricultural Pests & Disease Act, 1954
- The Acquisition of Land for Flood Control and Prevention of Erosion Act, 1955
- The Bihar Control of the Use and play of Loudspeakers Act, 1955
- The River Boards Act, 1956
- The Ancient Monuments and Archeological Sites and Remains Act, 1958
- The Kerala Agricultural Pests & Disease Act, 1958
- The Atomic Energy Act, 1962
- The Major Port Trusts Act, 1963
- The Delhi Restriction of Land Uses Act, 1964
- The Beedi and Cigar Works Act, 1966
- The Insecticides Act, 1968
- The Maharashtra Water Pollution Prevention Act, 1969
- Tamil Nadu Water Supply and Drainage (Prevention & Control of Water Pollution. Act, 1970
- The Cattle Trespass Act, 1971
- The wild life (Protection. Act, 1972
- The Water (Prevention & Control of Pollution Act, 1974
- The Urban Land Act (Ceiling and Regulation, 1976
- The Madhya Pradesh Gandhi Basti Kshetra (Sudhar Tatha Nirmulan. Adiniyam, 1976
- The Territorial Waters, continental shelf, exclusive Economic Zone and other Maritime Zone Act, 1976
- The Water (Prevention and Control of Pollution Cess Act, 1977
- The Water (Prevention and Control of Pollution Amendment Act, 1978
- The Coast Guard Act, 1978

- The Forest (Conservation Act, 1980
- The Air (Prevention and Control of Pollution Act, 1981
- The Fairways Act, 1981
- The Narcotic Drugs and Psychotropic Substances Act, 1985
- The Environment (Protection Act, 1986
- Motor Vehicles Act, 1988
- Hazarduous Wastes (Management and Handling Rules, 1989
- Manufacture, Storage and Import of Hazarduous Chemical Rules, 1989
- The Public Liability Insurance Act, 1991
- The Public Liability Insurance Rules, 1991
- The Wild Life (Protection, Rules, 1995
- The National Environment Appellate Authority Act, 1997

Annexure - 2

Central Legislations

Air Pollution

- The Indian Boilers Act, 1923
- Motor Vehicle Act, 1939 (Repealed by Act No.59 of 1988.
- The Mines and Minerals (Regulation and Development) Act, 1947
- The Factories Act, 1948
- The Industries (Development and Regulation. Act, 1951
- The Air (Prevention and Control of Pollution. Act, 1981 (amended in 1987.

Water Pollution

- The River Boards Act, 1956
- The Merchant Shipping (Amendment) Act, 1970
- The Water (Prevention and Control of Pollution. Act, 1974 (amended in 1988.
- The Water (Prevention and Contol of Pollution. Cess Act, 1977 (amended in 1991.

Radiation

- The Atomic Energy Act, 1962

Pesticides

- The Poison Act, 1919
- The Factories Act, 1948
- The Insecticides Act, 1968

Miscellaneous

- The Indian Fisheries Act, 1897
- The Indian Forest Act, 1927
- The Prevention of Food Adulteration Act, 1954
- The Ancient Monuments and Archaeological Sites and Remains Act, 1958
- The Wild Life (Protection. Act, 1972 (amended in 1983, 1986 and 1991)
- The Urban Land (Ceilings and Regulation. Act, 1976)
- The Forest (Consrvation. Act, 1980 (amended in 1988)
- The Environment (Protection. Act, 1986)

Annexure - 3

List of Projects Requiring Environmental Clearance from the Central Government

1. Nuclear Power and related projects such as Heavy Water Plants, Nuclear Fuel Compley, Rare Earths.
2. RiverValley projects including hydelpower, major irrigation and their combination including flood control.
3. Ports, Harbours, Airports (except minor ports and harbours).
4. Petroleum Refineries including crude and product pipelines.
5. Chemical Fertilizers (Nitrogenous and Phosphatic other than single super phosphate).
6. Pesticides (Technical).
7. Petrochemical complexes (Both Olefinic and aromatic) and Petro-chemical intermediates such as DMT, Caprolactam LAB etc. and production of basic plastics such as LLDPE, HDPE, PP, PVC.
8. Bulk drugs and pharmaceuticals.
9. Exploration for oil and gas and their production, transportation and storage.
10. Synthetic Rubber.
11. Asbestos and Asbestos products.
12. Hydrocyanic acid and its derivatives
13. (a) Primary metallurgical industries (such as production of Iron and Steel, Alunminium, Copper Zinc, Lead and Ferro Alloys) (b) Electric arc furnaces (Mini Steel Plants).
14. Chloralkali industry.
15. Integrated paint complex including manufacture of resins and basic raw materials required in the manufacture of paints.
16. Viscose Staple fibre and filament yam.
17. Storage batteries integrated with manufacture of oxides of lead and lead antimony alloys
18. All tourism projects between 200m - 500 meters of High Water Lin and at locations with an elevation of more than 1000 Meters with investment of more than Rs. 5 crores.

19. Thermal Power Plants.
20. Mining Projects (major minerals) with leases more than 5 hectares.
21. Highway Projects.
22. Tarred Roads in himalayan and or Forest areas.
23. Distilleries.
24. Raw Skins and Hides.
25. Pulp, Paper and newsprint,
26. Dyes.
27. Cement.
28. Foundries (individual).
29. Electroplating.

Annexure - 4

Industries which Require Industrial Licensing

1. Coal and Lignite
2. Petroleum (other than crude. and its distillation products.
3. Distillation and brewing of alcoholic drinks.
4. Sugar
5. Animal fats and oils and their preparations
6. Cigars and cigarettes of tobacco and manufactured tobacco substitutes.
7. Asbestos and asbestos-based products.
8. Plywood, decorative veneers and other wood based products such as particle board, medium density fibre board, and block board.
9. Leather
10. Tanned or dressed furskins.
11. Paper and Newsprint except bagasse based unit. (i.e. except units based on minimum 75% pulp from agricultural residues, bagasse and other non-conventional raw materials).
12. Electronic aerospace and defence equipment all types.
13. Industrial explosives including detonating fuses, safety fuses, gun powder, nitrocellulose and matches, explosives; pyrotechnic products; matches; pyrophoric alloys; certain combustible preparations.
14. Drugs and Pharmaceuticals (according to Drug Policy)
15. Entertainment electronics (VCRS, colour TVs, CD players, tape recorders).

Annexure - 5

List of Polluting Industries

1. Primary metallurgical producing industries viz. zinc, lead, copper, aluminium and steel.
2. Paper, pulp and newsprint
3. Pesticides/insecticides
4. Refineries
5. Fertilizers
6. Paints
7. Dyes
8. Leather tanning
9. Rayon
10. Sodium/potassium cyanide
11. Basic drugs
12. Foundry
13. Storage Batteries (lead acid type.
14. Acids/alkalies
15. Plastics
16. Rubber - synthetic
17. Cement
18. Asbestos
19. Fermentation industry
20. Electro-plating industry.

Annexure - 6

National Ambient Air Quality Standards (CPCB 1994)

Pollutant	Time	Concentration in ambient air		
		Industrial Area	*Residential Rural & Other Areas*	*Sensitive Areas*
Sulphur dioxide (SO_2)	Annual Average*	80 ug/m^3	60 ug/m^3	15 ug/m^3
	24 hours*	120 ug/m^3	80 ug/m^3	30 ug/m^3
Oxides of Nitrogen NO_x	Annual Average*	80 ug/m^3	60 ug/m^3	15 ug/m^3
	24 hours*	120 ug/m^3	80 ug/m^3	30 ug/m^3
Suspended Particulate Matter (SPM)	Annual Average*	360 ug/m^3	140 ug/m^3	70 ug/m^3
	24 hours*	500 ug/m^3	200 ug/m^3	100 ug/m^3
Respirable Particulate Matter (size less than 10 um) (RPM)	Annual Average*	120 ug/m^3	60 ug/m^3	50 ug/m^3
	24 hours*	150 ug/m^3	100 ug/m^3	75 ug/m^3
Lead (Pb)	Annual Average*	1.0 ug/m^3	0.75 ug/m^3	0.50 ug/m^3
	24 hours*	1.5 ug/m^3	1.00 ug/m^3	0.75 ug/m^3
Carbon Monoxide (CO)	8 hours*	5.0 ug/m^3	2 ug/m^3	1.0 ug/m^3
	1 hour*	10.0 ug/m^3	4.0 ug/m^3	2.0 ug/m^3

* Annual Arithmatic mean of minimum 104 measurements in a year taken twice a week 24 houly at uniform interval.

** 24 hously / 8 hourly values should be met 98% of the time in a year. However, 2% of the time, it may exceed but not on two consecutive days.

Annexure - 7

Bureau of Indian Standards for Drinking Water (IS: 10500-1983)

Sl. No.	*Substance or characteristics*	*Reuirement / desirable limits*
1.	Color, Hazen units, Max.	10
2.	Odour	Unobjectionable
3.	Taste	Agreeable
4.	Turbidity, NTU, Max.	10
5.	pH value	6.5 to 8.5
6.	Total Hardness (as $CaCO^3$) mg/1	300
7.	Calcium (as Ca) mg/1, Max.	75
8.	Magnesium (as Mg) mg/l, Max.	30
9.	Copper (asCu) mg/l, Max.	0.05
10.	Iron (as Fe) mg/l, Max.	0.3
11.	Magnese (as Mn) mg/l, Max.	0.1
12.	Chlorides (as Cl) mg/l Max.	250
13.	Sulphate (as SO_4) mg/l Max.	150
14.	Nitrate (as NO_3) mg/l, Max.	45
15.	Fluoride (as F) mg/l, Max.	0.6-1.2
16.	Phenolics (as C_6H_5OH) mg/l, Max.	0.001
17.	Mercury (as Hg) mg/l, Max.	0.001
18.	Cadcium (as Cd) mg/l Max.	0.01
19.	Selenium (as Se) mg/l, Max.	0.01
20.	Arsenic (as As) mg/l, Max.	0.05
21.	Cyanide (as CN) mg/l, Max.	0.05
22.	Lead (as Pb) mg/l, Max.	0.1
23.	Zinc (as Zn) mg/l, Max.	5.0
24.	Anionic Detergents (as MBAS) mg/l, Max.	0.2
25.	Chromium (as Cr^{6+}) mg/l, Max.	0.05
26.	Polynuclear Aromatic Hydrocarbons (as PAH) mg/l, Max.	—

Annexure - 8

CPCB Standards with Respect to Noise in Ambient Air

Area code	*Category of Area / Zone*	*Limits in dB(A) Leq**	
		Day time	*Night time*
(A)	Industrial area	75	70
(B)	Commercial area	65	55
(C)	Residential area	55	45
(D)	Silence zone	50	40

Note:

1. Day time is reckoned in between 6 a.m. and 9. p.m.
2. Night time is reckoned in between 9 p.m. and 6 a.m.
3. Silence zone is defined as areas upto 100 meters around such premises as hospitals, educational institutions and courts. The Silence Zones are to be declared by the Competent Authority Use of Vehicular horns, loudspeakers and bursting of crackers shall be banned in these zones.
4. Mixed categories of areas should be declared as one of the four above mentioned categories by the Competent Authority and the corresponding standards shall apply.

* dB(a) Leq denotes the time weighted average of the level of sound in decibels on scale A which is relatable to human hearing.

A "decibel" is a unit in which noise ie measured.

"A", in dB(A) Leq, denotes the frequency weighting in the measurement of noise and corresponds to frequency response characteristics of the human ear.

Leq: It is an energy mean of the noise level over a specifiec period.

Annexure - 9

State Pollution Control Boards/Pollution Control Committees of Union Territories

(Comunications to be addressed to
Chairman/Member-Secretary of the Board/Comimittee

State	*Address*
Andhra Pradesh	Andhra Pradesh Pollution Control Board, IInd Floor, HUDA Complex, Amerpret, Opp. Sarathi Studio
Arunachal Pradesh	Arunachal Pradesh Pollution Control Board, Govt of Arunachal Pradesh, Office of Principal Chief Conservator of Forests, Itanagar-791 111
Assam	Assam Pollution Control Board Bamubunaudabm, Guwahati-781021
Bihar	Bihar State Pollution Control Board, Beltron Bhavan, IInd Floor, Jawaharlal Nehru Marg, Shastri Nagar, Patna-800023
Goa (Panaji)	Goa State Pollution Control Board House No. 243, Patto, Panaji-403 001
Gujarat	Gujarat State Pollution Control Board, Sector 10-A, Gandhinagar-382043
Haryana	Haryana Pollution Control Board, Kothi No. 66 1, Sector 8-B , Chandigarh
Himachal Pradesh	H.P. Pollution Control Board, Hotel Kings, Top Floor, The Mall, Shimla-171001
Jammu & Kashmir	J&K Pollution Control Board, Dept. of Environment Ecology & Forests, Drug Research Institute Campus, Behind Govt. Silk Factory, Rajbagh,

	Srinagar-190008
Karnataka	Karnataka State Pollution Control Board, No. 25, 6th, 7th & 8th Floors, Public Utility Building, M.G. Road, Bangalore-560001
Kerala	Kerala State Pollution Control Board, Kettakayam Building, Plamoodu, Patta P. 0., Trivandrum-695001
Maharashtra	Maharashtra Pollution Control Board, Chhatrapati Shivaii Municipal Market Building, 4th Floor, Palton Road, Bombay-400001,
Madhya Pradesh	Madhya Pradesh Pradushan Nivaran Maiidal, Paryavaran Parisar, E-5, Arera colony, Bhopal- 462016
Meghalaya	Meghalaya Pollution Control Board, Arden, Motinagar, Shillong-793001
Mizoram	Mizoram State Pollution Control Board, Office of the Div. Forest Officer Working Plan Forests Division, C/0 Principal CC of Forests, Mizoram Aizwal-796001.
Manipur	Manipur Pollution Control Board, Urban Water Supply Circle, PHED, Khoyathong, Imphal-795001
Nagaland	Nagaland Pollution Control Board C/o Chief Secretary, Govt. of Nagaland Dept. of Forests, Ecology & Environment and Wildlife, Kohima, Nagaland-797001
Orissa	Orissa State Pollution Control Board A-1 18, Nilakantha Nagar Unit Vill, Bhubaneswar-751012

(Contd.)

State	*Address*
Punjab	Punjab Pollution Control Board Nabha Road, Patiala-147001
Rajasthan	Rajasthan Pollution Control Board J-2/35, Mahaveer Marg, C-Scheme Jaipur-302001
Tamil Nadu	Tamil Nadu Pollution Control Board, 32, Santhome High Road, Chennai-600004
Tripura	Tripura State Pollution Control Board, Kerchowmuhani, Krishna Nagar, Agartala, Tripura (West)-799001
Uttar Pradesh	U.P. Pollution Control Board Pragati Kendra, IInd Floor Kapoorthala Complex, Aliganj, Lucknow-226020.
West Bengal	West Bengal Pollution Control Board, 10, Camac Street, 2nd Floor, Industry House, Calcutta-700017

Annexure – 10

List of Regional Offices of the Ministry of Environment and Forests, Government of India

Sl. No.	*Regional Offices*	*Area*
1.	Chief Conservator of Forests(C) Regional Office, Kendriya Sadan, 4th Floor E&F Wings, 17th Main Road, II Block, Koramangala, Bangalore - 560034.	Andhra Pradesh, Goa, Karnataka, Kerala, Tamil Nadu, Pondicherry, Lakshadeep
2.	Chief Conservator of Forests(C) Regional Office (EZ) 194, Karvele Nagar, Bhubaneshwar - 751001	Orissa, Andaman & Nicobar Islands
3.	Chief Conservator of Forests(C) Regional Office (WZ) E-3/240, Arera Colony, Bhopal - 462016	Madhya Pradesh, Maharashtra, Gujarat, Dadar & NagerHaveli, Daman & Diu
4.	Chief Conservator of Forests(C) Regional Office (NEZ) Upland Road, Laitumkhrah, Shilong - 793003	Arunachal Pradesh, Assam, Manipur, Megalaya, Tirupara, Nagaland, Mozoram
5.	Chief Conservator of Forests(C) Regional Office (CZ) B-1172, Sector-K Aliganj, Lucknow - 226020	Uttar Pradesh, Rajasthan
6.	Chief Conservator of Forests(C) Regional Office (NZ) SCO 132, 133, Sector 34-A	Haryana, Punjab, Chandigarh, Himachal Pradesh, Jammu and Kashmir

5

Environmental Assessment

Introduction

Environmental assessment is a central but frequently underdeveloped component of comprehensive, multidimensional assessment with individual clients, families, and neighborhoods or communities. Social workers have always paid attention to environmental variables that may influence individual or group functioning, and nearly all definitions of social work assessment include generating information about the client's situation or environment. In direct practice, however, environmental assessment has focused primarily on the immediate psychosocial environment. The assumption appears to be that environmental data beyond the client's immediate environment is not centrally relevant to assessment in interpersonal helping. At the same time, the many useful organizing frameworks available for understanding the wider environmental contexts of community or neighborhood are designed primarily for generalist or community organization practice.

Our intention here is to bridge this gap between micro and macro approaches to assessment. We consider a comprehensive and announced understanding of our clients' environmental contexts — at every systems level — to be at the heart of effective, appropriate social work practice.

If social workers are to embrace environmental interventions on a footing equal to that of person-oriented interventions, they need to think and assess contextually. Studies of attribution bias in social work assessment suggest that social workers tend to view client issues as person-centered or psychological, rather than as situational or environmental. Subsequent interventions then tend to be focused on individual change and adaptation rather than on environmental change. As Kemp points out "faced with numerous depressed clients...a social worker in direct practice may feel more of a need to develop skills in treating depression than to think whether a community issue is contributing to his or her clients' problems".

In this chapter, we present a rationale and a range of strategies for conducting environmental assessments in social work practice settings. We agree with Jordan and Franklin that "the practitioner who separates the client from the environment risks losing the breadth and depth of the client's life experiences". In so doing, we stress that the primary goal of environmental assessment in person-environment practice is to empower the client to act in the environment, rather than to support worker treatment of environmental problems. We explore a number of assessment tools and techniques that allow both worker and client to gain a better understanding of environmental variables and their impact. We make a particular case for initiating environmental assessments at the level of personal social networks, for we believe that this level offers important opportunities for active consultation, collective exchanges, and empowerment. Furthermore, larger systems such as other human service organizations, neighborhood or community factors, or public policy become more accessible as targets of change when worker and client are armed with knowledge of the client's environment.

Assessment in Social Work

Accurate and timely assessment is critical to effective case planning and intervention. Social work assessments seek a systemic understanding of the case situation to guide practice decisions: a core practice task is to encompass relevant contextual features and to identify the interplay between the person and the impinging environment. During assessment, the client and worker gather, analyze, and synthesize information in the following general areas: (a) issues of concern, (b) coping capacities of clients and significant others, (c) relevant systems involved in issues of concern and the nature of reciprocal transactions between clients and these systems, (d) resources that are available or are needed to remedy the issues of concern, and (e) the client's motivation to work on issues of concern. Assessment in social work is best thought of as both a process and ac product. For both worker and client the process is essentially cognitive: information is gathered, meanings are ascribed, a framework for organizing information is applied, and implications are drawn for goals and interventions. As a product, assessment generally leads to an oral or written statement of the issues of concern, description of the client system, how the client functions in relation to other systems, and formulation of a plan for change. Depending on the practice setting, the assessment statement assumes many forms of varying length and complexity. The assessment framework applied in any given situation naturally flows from the agency setting and the responsibilities of the worker in this context. In general, however, social work assessments are focused on a consideration of the person and environment, encompass a multidimensional and multi-systems perspective, and are strengths-based, client-driven, and client-centered. Franklin and Jordan, for example, present an integrative skills assessment protocol, designed primarily for social work with individuals and families, which includes:

- information about the client,
- the nature of the presenting problem,
- intrapersonal issues (e.g., cognitive functioning, emotional functioning, behavioral functioning, medical information, mental status, ethnic/cultural considerations, motivation, roles and role performance, and developmental considerations),
- interpersonal issues (e.g., family composition and process, work/school relations, peer relations),
- agency context and social supports,
- strengths and resources at various systems levels.

Assessment In Person-environment Practice (P.E.P.)

As a product or outcome, assessment in P.E.P. contributes to understanding the issues confronting the client and is the foundation for the development of a collaborative plan of action. As a process, assessment involves the client or client system and worker as partners in the process of knowledge development and opens up new ways of looking at the world, contributing to the development of "critical consciousness", or transformed perspectives on person-environment interactions.

We described the process of empowerment and the critical role of the environment and defined environmental assessment as:

> *an ongoing process in which client and worker, in partnership, gather and critically analyze information on the client or client system in transaction with multiple levels of the environment, including strengths, resources, and opportunities, as well as risks, challenges, and issues of concern, and with attention to the meaning of these environmental experiences for the client.*

The environmental assessment methods presented in this chapter are central to empowerment practice for they allow for critical reflection and analysis of the environment through interaction and dialogue between client and worker and between client and others in similar circumstances. This process of meaning-making and consciousness-raising lays the foundation for effective planning and action.

Active client involvement in the generation and analysis of environmental information is the backbone of environmental assessment and intervention. In the assessment phase, the worker and client gather data about multiple levels of the environment, engage in a dialogue about the information gathered, and critically reflect on environmental impacts on the client's current situation. The very act of gathering information, asking questions about particular topics, and using strategies such as the visual display of information typically results in a shift in clients' perspectives on their situation. The process of gathering environmental information may change the client's definition of the presenting issue, toward more complex views of the interaction between personal and environmental factors. Careful environmental assessment is thus essential to the intervention or change process.

The notion here is that we are not merely completing an inventory of types and categories of environmental resources, but rather conducting a critical analysis with clients of their transactions with their environments. This process is supported by assessment methods that enable workers and clients to look at the clients' active responses to their environments and to realistically consider the opportunities and constraints in those environments, at various system levels, and in all their complexity.

How one defines the environment naturally sets boundaries on what factors are included in an environmental assessment. Our definition of environment includes both the objective environment and the many ways in which that environment is personally and socially constructed. Core dimensions of the environment include the physical environment, both natural and built; the social environment, largely comprising networks of human relationships at various levels of intimacy; the cultural, political, and institutional environment; and the perceived environment, as constructed in individual and collective systems of meaning and belief.

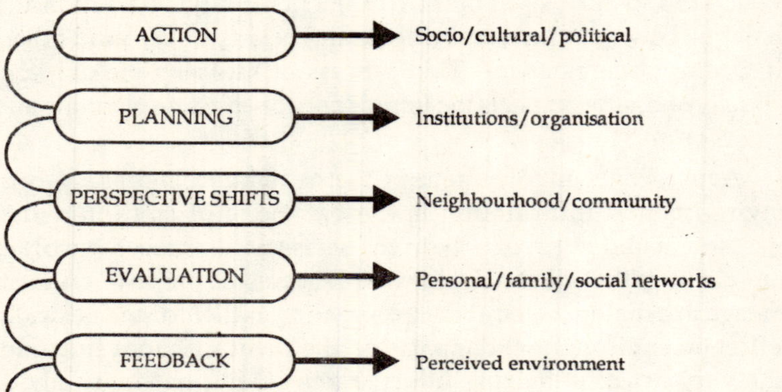

Fig. 5.1. Environmental assessment: partnership and process

"Environment" does not mean the same thing to everyone nor is it experienced in the same way. In Chapter 6 we discuss how factors such as race, gender, age, or ability affect one's experience and construction of meaning at various levels of the environment. Environmental assessments must tap into individual and group experiences with the environment and the meanings these hold for our clients and incorporate the realization that the environment we see as workers may not be the environment our clients experience. Consider, for example, the differential impact of physical settings:

> *When a flight of stairs leads to the entryway of a building, each person entering that building must alter his or her gait to accommodate the change in surroundings. However, while the weekend jogger may barely notice this shift and experiences no feeling of impediment, someone walking with the assistance of a cane may experience the stairs as an obstacle to be negotiated cautiously.*

A person's social and class position influences access to opportunities and power in the environment. This can be assessed through a power or stratification analysis. A power analysis involves assessment of the ways in which conditions of powerlessness (based on race, age, social class, gender, sexual orientation, etc.) contribute to the current situation faced by the client. The analysis includes identifying sources of both power and emancipation, including sources that may have been neglected or underutilized, such as social network resources.

Environmental assessments should not be overwhelmingly focused on environmental threats and deficits, but should attend equally to the strengths and resources, in individuals and communities, that form the basis for new connections and relationships. We noted that the strengths perspective represents a shift in social work focus consistent with empowerment and environmentally based intervention. In some areas of social work practice, however, such as the mental health field, the historical tradition has been to overemphasize the negative influences of the environment on people. A focus on environmental deficits may unnecessarily narrow our vision of helping resources and restrict client participation in the community. Within the community mental health system, for example, failure to recognize, recruit, and utilize naturally occurring environmental strengths results in further exclusion and social isolation of persons with chronic mental illness.

Sources of Data for Environmental Assessment

As with all aspects of the assessment process, environmental assessment is strengthened if data are generated using a variety of assessment methods and drawing on multiple perspectives. As we employ various assessment methods, such as client self-report and participant observation, and explore more than one level of the environment, our environmental assessments become more complete and valid. Other social work assessment methods may include standardized interviews with clients, significant others, and "collaterals" (others who interact with the client); ethnographic interviews with clients, significant others, and collaterals; background information sheets and questionnaires; psychological, projective, or standardized assessment tests; client logs or diaries; and behavioral observations. Typically, graphic and mapping techniques, such as an ecomap or genogram, are used to organize and display some of this information. Computer-assisted assessment packages have also emerged as promising tools in the assessment process. In P.E.P., priority is given to methods that most fully allow for the active involvement of the client in the generation of data and for discussion and reflection with the worker or others, as these methods support the goal of empowering environmental intervention.

In work with a community group, for example, environmental assessment techniques might include review of printed materials about the community, such as newspapers, maps, and newsletters; formal and informal interviews with key informants, residents, and other community stakeholders; interaction and participation with small groups and organizations within the community; and careful participant observation of community events and interactions. Much can be learned about a community from printed sources and other regularly collected demographic information, but physically driving or walking through a community gives the worker a firsthand sense of key factors such as physical surroundings, both natural and built, obvious geographical boundaries, and general impressions of safety. Talking with major community figures likewise provides greater

insight into residents' perceptions of themselves and their community. Warren stressed the importance of interviewing the elementary school principal, for example — someone who is in regular contact with families and is knowledgeable about the economic situation, racial tensions, and other issues in the community. Interviews with community residents also provide important information on residents' definitions of neighborhood or community boundaries, which may bear little or no resemblance to official census tracts or planning units.

Direct contact with informal, local sources of information is particularly important in work with communities of color. Dialogue with community members provides an understanding of cultural, political, and social factors and information about both differences and alliances among various groups within the community (e.g., religious, ethnic, racial, educational, social class groupings) as well as shared goals.

The type and amount of information to be collected in the environmental assessment will depend on many factors, including the client's situation and commitment to involvement, the type of assessment tool or method selected, the purpose of social work involvement, and the agency's mission and goal. In child protective work, for example, risk assessment tools routinely screen for social and physical environmental factors that may place the young child or youth at greater risk. On the other hand, hospital discharge planners working with elderly stroke victims for their return home may assess a very different set of environmental factors. In each instance, the age and condition of the client, the immediacy of the need for service, relevant cultural features, and the agency's overall service mission will influence the choice and depth of assessment information collected.

Frameworks for Environmental Assessment

Common sense tells us that environmental assessments should attend to those aspects of the environment most relevant to the client's situation. However, we have also seen that, depending on factors such as age, cultural background, and class, individuals experience the environment differently. Ultimately the fit between the person and the environment is the central focus of assessment.

The environments of two clients may provide vastly different resources, for example, but for each the environment may be sufficient for present needs and interests. Conversely, two clients may share the same environment but experience it in very different ways. These dramatic differences in how people perceive, experience, and utilize resources underline the central importance of individualization in the assessment and intervention process.

Aspects of the Environment

Aspects of the environment that should routinely be screened in any assessment include (a) the extent to which basic human needs are being met by the environment, (b) strengths and resources present in the environment, and (c) environmental toxins and hazards.

Does the Environment Meet Basic Needs?

Although the social work profession has always been concerned about the physical environment, many social workers proceed with intervention as though it does not really matter. Much credit for changing this perspective is due to the home-based and family support movement. Practitioners working in home and community settings, who are deeply involved in the physical environment and essentially become part of the client's milieu, simply cannot ignore it. Teaching behavior management skills to clients in an office setting is one thing; it is quite another to help them implement behavioral changes when rats are dashing in and out, sounds and smells assault, and there is no clear wall (or refrigerator) upon which to hang a behavior chart. Likewise the assessment of concrete needs is often the first priority in working with clients in home and community settings.

Hepworth and Larsen identify a number of universally needed social and physical environmental resources. These range from the immediate home environment (e.g., ample space, sanitation, food) to the surrounding community (e.g., adequate police and fire protection). Zastrow suggests a similar list of basic environmental needs, with the addition of safety from environmental hazards and safety from air, noise, and water pollution. Zastrow also identifies some specific social support needs, such as emotional support from significant others and support from significant others to be drug-free, as basic universal needs. Vosler, Dunst and Dunst also provide assessment tools to determine the extent to which basic resources are available, accessible, and adequate to meet perceived needs, as well as the extent to which needs that are not adequately met contribute to stress and strain.

What Are the Environment's Strengths and Resources?

Despite considerable recent attention to the strengths perspective by the social work profession, the components of a personal strengths assessment have been articulated more fully than those of an environmental strengths assessment . As with social work assessments in general, a strengths assessment should be multidimensional in nature, focusing both on internal and external strengths. In general, environmental strengths have been conceptualized as consisting of availability of and access to resources in the social environment, such as support from family, friends, or relatives.

Table 5.1. Universally Needed Environmental Resources

- Adequate social support systems
- Access to health care
- Access to day care services
- Access to recreational facilities
- Mobility to socialize, utilize resources, and exercise rights as citizen
- Adequate housing that provides ample space, sanitation, privacy, and safety
- Adequate police and fire protection
- Safe and healthful work conditions
- Adequate financial resources to purchase essential resources
- Adequate nutritional intake
- Predictable living arrangements with caring others
- Opportunities for education and self-fulfillment
- Access to legal resources
- Access to religious organizations
- Employment opportunities

One useful model for depicting environmental versus personal strengths is Cowger assessment framework, which consists of four quadrants revolving around two axes. The horizontal axis represents environmental factors versus personal factors as contributors to the need for service, and the vertical axis represents a strengths versus deficits continuum. Using this model, a complete assessment will ideally lead to sufficient understanding of both environmental and personal strengths and issues of concern.

Franklin and Jordan identify potential environmental strengths and resources as comprising availability of employment, adequate housing, supportive family, network of friends, community ties, religious network support, and financial security. Pecora and English provide guidelines for assessing environmental strengths within multicultural communities. In their review of risk and protective factors influencing child maltreatment, a number of community strengths are identified and operationalized, including community cohesiveness, community racial identity, community resources for children, community resources for adults, community groups and organizations, and community interaction with mainstream agencies.

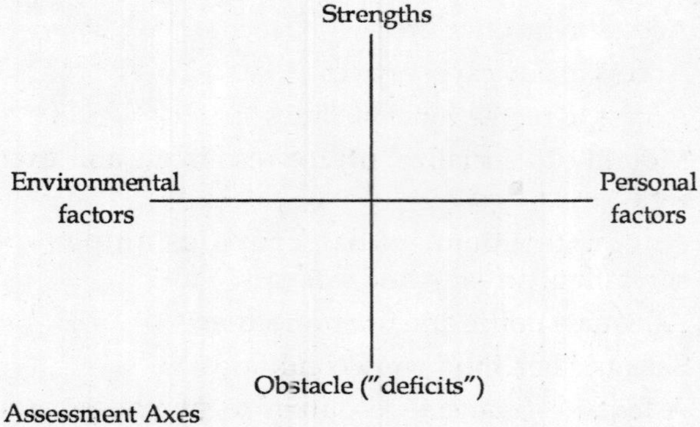

Fig. 5.2.

It is important to remember that an assessment of environmental strengths and resources need not and should not be limited to tangible, physical aspects of the community. Attitudinal and experiential factors, such as community identity, degree of attachment to the community, and shared community values, as well as relational factors, such as the community's ability to relate to other groups and institutions, are important domains of a strengths-based assessment. Finally, cultural narratives and folklore are extremely important sources of strength. These may include cultural accounts of origins, migrations, and survival, which can bring deep meaning and inspiration for change both to individuals and groups.

Are Environmental Hazards Present?

Environmental hazards include substances and conditions in the environment that are known or suspected to cause or increase the risk of physical, mental, or social impairment. In recent years, newspaper accounts and television news stories have been filled with reports of the effects of contaminated air, water, and food, hazards in the workplace and home (e.g., carbon monoxide), radiation exposure, noise, crowding, and dangerous consumer products. Though social workers frequently work with victims of environmental hazards (e.g., people with health problems related to exposure to asbestos or exposure to other occupation-related substances; veterans exposed to chemicals used during the Vietnam and perhaps Persian Gulf wars), our assessment tools and procedures rarely do justice to their impact. Consequently, many social workers are unaware or ill-informed about these and other environmental hazards. While the direct and indirect effects of many substances are still unknown or difficult to predict, there is growing empirical evidence to warrant further consideration of environmental hazards as part of a complete assessment. Consider for example some of the following examples of threats to physical and mental health from environmental pollution:

- The link between carbon monoxide levels and depression, among other aspects of cognitive and emotional functioning,
- the relationship between lead poisoning and mental retardation and other behavioral and developmental problems of children,
- the rise in the incidence of asthma and other respiratory diseases in response to air pollution,
- fetal deaths, miscarriages, and birth defects associated with exposure to toxic wastes in water and landfills.

Environmental equity is a growing concern within the social work profession. Risk from environmental exposure to hazards is not shared equally but is borne disproportionately by low-income and minority communities: the racial makeup of the community, for example, is a strong predictor of toxic waste site location; poor neighborhoods and communities are at greater risk of industrial accidents; and many indigenous groups are at increased risk due to use of their natural lands and resources as sites for mining or power production. Taken as a whole, the evidence strongly points to the need to ask about and consider environmental hazards when working with vulnerable populations.

Levels of Environment

Jordan and Franklin provide a summary of significant environmental assessment factors at various levels of the environment that are consistent with the model of P.E.P. presented here. First is the family/significant other environment, which includes family of origin and present family constellation. The next level of environment to be considered is the social network of friends, including the quantity and quality of social support provided. The quality of the work or school environment is an important level as well, particularly in relation to work/school history, work/school environment (e.g., stressful, crowded, or dangerous conditions), and work/school relationships. The neighborhood and community environment is yet another assessment consideration; the adequacy of living conditions and the nature of the client's relationship with the community are examples of factors at this level. Finally, broader governmental and social forces are the last level to assess in conducting a person in environment assessment. Considered here should be the "relative economic and social state of affairs" and broader social forces that may be impacting access to or availability of basic social services. As Jordan and Franklin point out, some clients (persons of color, ex-prisoners, persons with mental illness, persons with AIDS) may lack access to certain community resources due to discrimination.

Classification Tools

Several useful instruments provide access to the various levels of the environment (family, neighborhood, community) and enable more careful assessment of environmental impacts. We summarize many of these environmental assessment instruments, tools, and techniques at the end of this chapter.

Person-in-Environment (PIE) System

The person-in-environment (PIE) system is a four-factor classification system designed by social workers to complement the classification system used by the psychiatry profession, but focusing instead on social and environmental factors. Four factors are used to describe a client's problem situation: Factor I identifies problems in social functioning; Factor II describes problems emanating from the environment; Factor III describes mental health problems; and Factor IV lists physical health problems. This system provides a means to describe and code social functioning of adults in terms of role performance in the various contexts of the environment. Six categories of environment are included in Factor II of this system: economic/basic needs (e.g., food, shelter, employment, economic resources, transportation), education/training, judicial/legal system, health safety and social service system, voluntary associations (e.g., religion, community groups), and affectional support system. The worker manual accompanying the PIE system lists several problems that may be experienced under each category of environment. The presence of discrimination in the environment is also a coded problem category, along with ratings of severity and duration. This system is currently awaiting large-scale tests of reliability and validity. While PIE is increasingly widely disseminated, this assessment system has some significant limitations for social work practice. These include a focus only on individual adult functioning, a focus on discrete problems, the requirement that PIE be used in conjunction with a psychiatric diagnosis, and the system's reliance on lists of factors rather than a more transactional approach.

Ecomap

Perhaps the best-known and most frequently used environmental assessment technique, the ecomap generates information on three major aspects of environmental resources: those used by the client; those available but not fully used by the client;

and gaps in resources, that is, those needed by the client but not present or accessible in the community. The ecomap is a useful method for visually documenting the client's relationship with the outside world, including the flow of energy and the nature of relationships experienced. Developed initially in the adoption practice arena, the ecomap has since been applied and adapted in a large variety of practice settings.

To complete an ecomap, clients write and encircle the names of people, groups, and organizations in their social networks. The distance between circles indicates closeness of relationships. Lines drawn between the circles represent the quality of the relationship: stressful, tenuous, positive. Arrows indicate reciprocity, the give and take of the relationship. Ecomaps are generally constructed jointly by worker and client, providing an opportunity for each to understand better the social context of the client's life. The basic format can readily be adapted to fit a variety of clients and settings.

Major Outcomes by Level of Environment

At various points in the process of conducting an environmental assessment with a client or a client group, both the worker and the client will have information to guide the selection of a target goal and the implementation of an intervention plan. In general, answers to the following questions should be available in usable form after completion of the environmental assessment.

Overall Environment

- What kinds of environmental strengths and resources are currently used by the client?
- What is the nature of the relationship between the client system and the environment?
- What strengths and resources in the environment are underutilized but could be mobilized?
- What resources are needed or required that are lacking in the environment?
- What obstacles (internal and external) prevent access to environmental resources?
- To what extent does the client have an enhanced understanding of the environment?

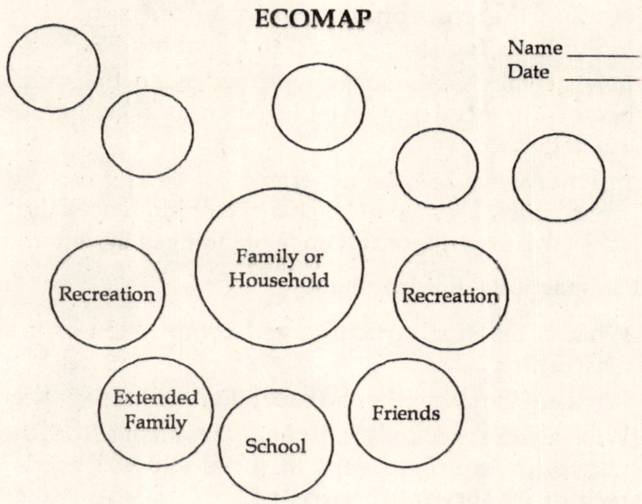

Fig. 5.3.

Physical Environment

To what extent does the physical environment provide (or is perceived as providing)

- Safety, security and shelter?
- Social contact?
- pleasure and recreation?
- support for completion of developmental and life tasks?
- adequate stimulation for growth? To what extent are the people connected with the physical environment
- dependent on that environment?
- lacking in power, skills, or resources to change the environment?
- involved in decision-making about environmental changes?
- involved in personal or collective action to achieve change in the environment?
- exposed to environmental disparities and risks as compared with other groups of people living in different environments?

Perceived Environment

- How is the environment experienced subjectively by individuals, families, groups, and communities?
- In what ways has the subjective experience of the environment been influenced by variables such as race, gender, and ethnicity?
- In what ways has the experience of being involved in an environmental assessment changed or influenced the client's subjective experience and understanding of the environment?

Social/Interactional Environment

- What is the size, structure, and composition of the social network?
- What are the strengths and capabilities of the social network?
- What are the types of social support available from the social network? Are there gaps in social support needs? Does reciprocity appear to be an issue?
- Which network members are identified as responsive, effective, accessible, and dependable?
- Which network members are critical of the client?
- What obstacles (internal and external) prevent the client's utilizing social network resources?
- How are social support needs prioritized in relation to other presenting issues or needs? Is crisis intervention an issue?
- What would be a viable social support goal? What would constitute an appropriate social network intervention?
- Within a family or group, how are individual members differentially connected to the social environment?

Institutional/Organizational Environment

- What has been the nature of relationships and the place of larger systems in the client system's life to date?
- What new relationships can be developed between the client system and the larger institutional system? How can viable relationships be maintained?
- What constraints or barriers (internal and external) limit or prevent the development of viable relationships with larger systems?

An important set of environmental assessment questions relates to the agency context in which much of social work practice takes place. Here, it is important for the worker to consider how the agency is perceived by the client and the community, and how those perceptions may or may not influence the social work relationship and task. For example, in what ways does the physical structure of the agency influence the work with clients? In some settings, group or family sessions are discouraged due to lack of space. How is the agency perceived in the community: as a helpful resource or as an "alien" presence? Does the agency have a good or bad reputation, and on what basis was this reputation formed? How are staff and workers supported or not supported by the agency? Are workers frustrated by activities that do not appear to be directly related to meeting client needs, and how does this affect their practice? Home-based intervention programs, for example, have found it useful to attend to worker needs for support and recognition, inasmuch as their work is often stressful, demanding, and isolating. Assessment of the agency environment can take place formally, as in convening focus groups or conducting client surveys. Information about the client's perceptions of the agency may also emerge from social network data. In any event, the impact of the agency setting on the worker and client needs to be considered.

Social/Political/Cultural Environment

- What is the nature of the sociopolitical and economic environment?
- To what extent are oppression and discrimination dominant features of transactions with the environment?
- To what extent does the immediate environment reinforce wider social patterns of power, privilege, or access to resources?
- What opportunities exist for sharing/validating cultural, ethnic or religious meanings and values?

Environmental Assessment and Stage of Helping

Assessment is often viewed and experienced as an ongoing process, changing in response to the various phases of helping. As new data emerge in the life of a "case" and as the helping relationship evolves, the assessment statement is modified and enlarged. Frequently, the situation is most completely understood during the time period when worker and client are actively involved in implementing interventions. For this reason, when worker and client are stumped or slowed in their progress, it is often assessment to which they return. Although we cannot fully cover each stage here, it is helpful to be reminded of the role of environmental assessment in various stages of helping.

Environmental assessments can serve useful functions in the early stages of helping when worker and client are just beginning to know one another. For example, environmental assessment at intake helps identify the need for concrete services or referral to other resources. Environmental assessment techniques that fully involve clients and provide them with immediately useful information about their situation are particularly helpful in supporting engagement and the development of client-worker partnerships. We have found that use of a social network map early on in the working relationship fulfills the functions both of assessment and engagement. Clients learn more about their social network resources in a non-threatening manner, and workers learn who else in the client environment should be included in ongoing assessments.

Environmental assessment also contributes to intervention planning, in that aspects of the environment that currently maintain the behavior of concern as well as potentially available resources to support behavior changes can be identified and mobilized as needed. For example, in previous work we examined social network interventions across the continuum of child welfare services and outlined the types of social network assessments that should be an integral part of each stage in the continuum from pre-placement to aftercare services. Environmental assessment questions to be considered in working with families of children in placement include:

- Will the family's network reinforce the parents' behavior changes or will the network sabotage behavior change efforts in unanticipated ways?
- Will the family's network be able to play a role in supporting parental visitation, either through emotional support or through concrete assistance in providing transportation to and from visits?

- Are there culturally significant network members, such as grandparents, with whom the child can maintain contact and connection even if contact with biological parents is limited or impossible?

During the intervention stage, it is important to ensure that all the relevant people and environmental systems in the client's life situation are involved to the extent required to achieve the goal. Family group conferencing, a technique that is described more fully in the next chapter, is one strategy that both assesses the environment and makes full use of that environment in intervention. The extent to which the client has changed but the environment has remained unchanged is also a critical question. Rzepnicki emphasizes the absolute centrality of environmental supports and conditions to the maintenance of intervention gains. In a review of the literature on factors that increase the likelihood that progress will be maintained after formal intervention ends, she suggests that workers should (1) create as much similarity as possible between the intervention situation and the client's natural environment, (2) increase environmental support for improved client functioning, including both social support (from natural networks and helpers) and concrete resources (both currently and in the future), and (3) provide multiple opportunities for learning of new skills and competencies. At closure or termination, environmental assessment and intervention are thus crucial in guiding the transition from reliance on the helping relationship with a worker or agency to connections with other sources of formal and informal community support.

Mapping Social Network Resources: The Social Network Map

We have found that social network assessment is often the best and most time-effective method for beginning an environmental assessment.

Gathering information at the level of the social environment naturally opens a window of opportunity to learn about other levels of the environment. This is because the type and quality of the social network and social support resources available to a client are heavily influenced by other environmental variables, including physical features, institutional structures, and social/cultural/political systems. For example, Cochran and Niego, in a review of parenting and social networks, aptly point out that factors such as "poverty, unemployment, and lack of educational opportunity dramatically influence the networks of families, limiting their capacity to nurture and support parents and their children".

A social network assessment can also be used to identify the different types of communities with which an individual maintains connections and the quality of those relationships as perceived by the individual. Community membership is based not only on residence, but on lifestyle, employment, interest, or experiences. Rubin identify five major types of communities: neighborhoods (the most commonly used definition), solidarity communities, social classes, networks, and communities of interest. Kemp points out that "some clients who seem relatively isolated in their local neighborhoods may have connections with wider communities of interest".

If social workers are to work with, and not against, the client's social network, completing an accurate assessment of social network resources is a necessary first task. This is especially true if, as is often the case, a large percentage of helping exchanges are occurring within the informal helping network. All too often, important social resources are not identified at all or are identified too late to be of any help in implementing a case plan.

Cultural differences between the client and the worker may exacerbate this difficulty; a social network perspective enables the worker to understand culturally specific patterns of help giving and receiving. The process of family group conferencing, for example, was informed by the traditional help-giving patterns of the indigenous peoples of New Zealand; this model was adopted from that group's definition of and preference for extended family involvement in decision-making. An understanding of coping and social support strategies used by various communities of color allows for more effective and early positive involvement of available resources.

There is also some evidence that naturally occurring sources of social support can be supportive of professionally based intervention with clients, thus allowing for more flexible and strategic use of time. In one study of network therapy, in which workers met the mental health client's entire social network, network therapy participants showed a 76 percent decrease in formal service contacts as compared with a 17 percent decrease in formal service usage for the comparison group. Similar findings emerged from a social network case management project with homeless women. When a support system is activated, the worker's role can become progressively decentralized. The decrease in service usage may be due to the fact that network members help detect and intervene with pending crises and help directly during periods of needed crisis intervention; in this manner, help provided from networks may be considered a factor in reducing phone calls to or face to-face contacts with case managers.

As a research technique, social network mapping has been applied with many client populations. Network analysis has been used with personal networks as well as interorganizational networks for service delivery. A number of tools are available for measuring such variables as social embeddedness, perceived social support, and enacted social support. The social network mapping tool that we have developed was drawn primarily from the work of Fraser and Hawkins and Lovell and Hawkins, with substance abusers and maltreating mothers, respectively. Whereas these researchers used the technique of gathering network member data and then developing a matrix to gather data on network member qualities and relationships, the current social network map enables collection of information on the total size and composition of the network and the nature of relationships within the network as perceived by the person completing the map.

Steps in Administration

Administering the social network map involves listing network members from each of seven domains (household, family/relatives, friends, people from work or school, people from clubs, organizations or religious groups, neighbors, and agencies or other formal service providers). Names or initials are initially recorded on the circle map and listed separately on small slips of paper or index cards. The next step involves asking a series of questions regarding the nature of network relationships. These questions cover (a) the types of support perceived to be available and (b) the extent to which network relationships are reciprocal, the extent to which network members are critical of the client, the closeness of relationships, the frequency of contact, and the length of relationships. The client is asked to sort the slips of paper or index cards into the appropriate response category (e.g., hardly ever, sometimes, almost always) using a sheet of paper as a sorting card. This procedure was informed by the work of Lovell, who used such an approach with maltreating parents. Number-coded responses to these questions are then recorded on the network grid.

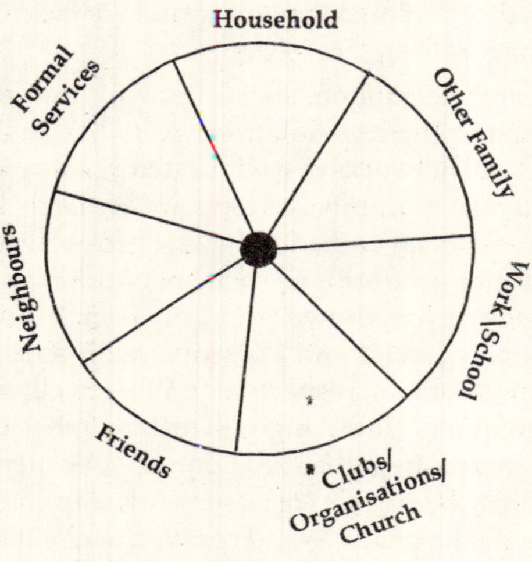

Fig. 5.4. Social network map

When analyzed, the social network map provides visual information as well as absolute numbers and percentages along the following dimensions:

- *Network size*: total number of people identified in the network.
- *Domain size*: total number/proportions of people in each of the seven domains.
- *Perceived availability of emotional, concrete, and informational support*: proportion of network members rated as "almost always" available to provide these types of support.
- *Criticalness*: proportion of network perceived to be "almost always" critical.

- *Closeness*: proportion of network perceived to be "very close."
- *Reciprocity*: proportion of network relationships in which "help goes both ways."
- *Directionality*: proportion of network relationships in which help goes primarily from client to network and proportion of network relationships in which help goes primarily from network to client.
- *Stability*: length of relationships (how long known).
- *Frequency of contact*: how often seen.

In addition to these specific variables, the social network map, when combined with other assessment information, may help the worker:

- Understand culturally specific patterns of help-giving; culturally relevant definitions of family, and culturally specific help-giving roles;
- Identify potential sources of support to aid in the maintenance of intervention gains;
- Understand family system boundaries;
- Pinpoint sources of conflict within the personal social network;
- Appreciate the client's perception of support;
- Encourage the client to actively restructure the immediate social environment;
- Encourage client efforts at self-help;
- Identify patterns of reciprocal helping;
- Identify relevant others who may participate in future network interventions.

Information on the reliability and validity of social network map data is often difficult to determine. Problems of recall, fatigue, and recent events all affect self-reports of social networks. In one test-retest study of the social network map's reliability, respondents identified 70 percent of network members when the map was administered a second time; ratings of network members were 76 percent reliable. Evidence suggests that some relational aspects of social networks, such as criticalness, were less stable than others. In another study, we attempted to address the issue of convergent construct validity, that is, correlation among dimensions of perceived social support as measured by the social network map and by other standardized social support instruments. Greater satisfaction with emotional support was associated with greater support satisfaction overall. On the other hand, the frequency of criticism from network members was significantly associated with less social support satisfaction. This study also confirmed our earlier findings that information about structural features alone, such as network size, provided little or no information about the quality, amount, or experience of social support. In general, further studies of reliability and validity of social network data are needed, particularly those which include shorter test-retest intervals, verification of reciprocated network membership, and measures of received support in addition to perceived support.

The Social Network Map in Practice

We have completed several pilot studies of the use of the social network map, in a home-based family service program in a large youth-serving organization and within Head Start early intervention. The clinical usefulness of the map in each of these settings was highly rated by participating staff: In all these contexts, the social network map also provides a significant baseline of environmental information. The map helped practitioners to systematically identify and assess stressors, strains, and resources within the client's social environment. Rather than describing social support in global terms (e.g., extended family lives nearby), the map provided specific information both to the workers and clients in areas such as the types of support available, the presence of close relationships, and the direction of help. Because of its general applicability, the social network map has been cited in numerous social work practice texts as a viable means of assessing client support systems.

The social network map has also been adapted for use in a number of practice settings and within different cultures. The map and accompanying instructions have been translated into the spoken languages of targeted client groups, a process that requires careful attention to meaning and nuance. Even when English is the spoken language, words may hold different connotations within different cultural groups. It is important to use words and phrases that will be understandable to the people who will use the map. Some practitioners have made adaptations in the titles of the social network domains so as to be more meaningful to particular client groups. For example, one project in Canada, which translated the map into French and Inuktitut, used the title Elders rather than Clubs/Organizations/Church when using the map with native populations, and referred to the social network map as People and Groups in Your Life. Others have found creative ways to use the social network map with all the members of a family, engaging adoptive families, for example, in a problem-solving process to decide as a group who were the most important members of their social network. The social network map has also been used to assess support systems for kinship foster care providers and as a research tool in studies of child maltreatment.

The social network map is a versatile and flexible clinical assessment tool that helps to identify potentially useful resources and the client's perceptions of those resources. Often this process leads to the discovery of resources that are available but underutilized. For example, one client who spoke negatively of her neighbors realized after completing the social network map that those same neighbors did provide her with specific types of support. Another woman in desperate need of child care was able to identify two potential people to help with baby-sitting and, with the worker's coaching, was able to enlist these people and establish a child care schedule. One client realized in working on the map that she had become isolated due to a recent move; a specific intervention was developed to initiate contacts with her new neighbors. The case vignettes at the conclusion of this chapter illustrate the ways in which social network data are considered in relation to the presenting problems and needs of the client. Although there are variations depending on the type and orientation of social service agency, information gathered from the network map in all cases directly informs the goals and interventive techniques established with the client.

The social network map is also an empowerment tool, for it enables those completing the instrument to gain insight into and ultimately more control over their social environments. Completing the map gives people the opportunity to review current social ties, reconsider past social ties, and anticipate future social ties. This knowledge, in turn, often leads to an increased sense of competence and control. In fact, one proposed model of the mediating effects of social support in the face of life stressors suggests that perception of social support may prevent a stress appraisal response in the face of a stressor. In other words, the perception that there are others to help leads one to a redefinition of the event or situation as less stressful. Social network members may also provide a solution or coping behaviors in response to a stressful event. As an intervention example of this principle, a single father was provided a copy of his social network map to hang near his telephone, to remind him who might be available to help in an emergency. The simple act of learning about one's past and present resources is an empowering experience, either because these resources become more immediately accessible, or the intervention required to make them accessible becomes more apparent. In one of our pilot projects, for example, the parents of a ten-year-old boy with a developmental disability were taught ways of establishing more productive relationships with the child's school staff; this in turn, unleashed sorely needed resources from which they had been previously alienated.

The experience of social network review is not always pleasant. In some instances, it becomes painfully clear that the current social network will not or cannot, even with help, support the changes desired or needed by the client. It may become clear in the network review, for example, that friends who are drug users will undermine progress during recovery. In such instances, a decision may need to be made to extricate oneself from a particular network or set of network ties and enlist new network members. On the other hand, a review of past network resources may lead to a plan to reconnect with these relationships as a means of generating renewed sources of support.

Before we leave this section on the social network map in practice, there are a few additional practical matters to address regarding its administration. Completion of the social network map involves filling in the circle map and the accompanying grid. The circle map can be enlarged, if necessary, to allow for more writing room. We have also used concentric circles within the map to depict closeness of relationships. In classroom and workshop exercises involving the map, we have routinely asked participants to draw arrows to indicate the flow of help and to use symbols directly on the map to illustrate the type of help provided by that source. Whereas the map can accommodate any number of network members, the grid has spaces for only fifteen names. We have typically asked people to list the top fifteen members of their network. However, it is not necessary to restrict the network in that manner. We suggest that another grid sheet be added to accommodate larger networks. Finally, it is not uncommon for people to list unexpected supports, such as deceased network members, religious figures (e.g., Jesus), or pets, on social network maps. Our approach has been to list these supports on the circle map exactly as they are described to us. We have often created another circle for deceased network members in the recognition that the relationships these figures represent can influence the decisions people make and how they feel about themselves. Similarly, we recognize that various forms of spirituality serve important social support and coping functions. We believe that by not "editing out" or applying our own framework of thinking on the information provided, we reach a more informed and ecologically valid picture of the client's social environment.

Guidelines for Social Network/Support Assessments

The following questions are helpful in translating information generated from the social network map into appropriate social support change goals:

1. Who is in the network, how are they related to the client, and who could be potential members?
2. What are the strengths and capabilities of the social network? Among the strengths that need to be examined are number of supportive relationships, variety of supportive relationships, types of support available (emotional, concrete, and informational), and reciprocity among helping relationships. Again, there is no one perfect network: only a network that meets the client's need.
3. What are the gaps in social support needs? Is there a poor fit between the types of support that network members are willing or able to provide and the types of support the client needs or desires?

4. What relationships in the network are based on mutual exchange? Does reciprocity seem to be an issue for the client? Is the client always giving to others and thereby experiencing stress and drain? Or does the client appear to be a drain on the network, with the result that network members are stressed and overburdened?
5. What network members does the client identify as responsive to requests for help, as effective in their helping, and as accessible and dependable? Do sufficient numbers of network members meet these Conditions? These are the people and resources that would most likely be an asset to any intervention plan.
6. What network members criticize the client in a negative or demanding way? Is the client surrounded by a network that is perceived as negative, non-supportive, or stress-producing?
7. What barriers exist to utilizing social network resources? Does the client lack supportive resources or lack skills to gain access to resources? For example, the client may lack some key social skills such as reaching out to others, offering feedback, requesting help, saying thank you when appropriate, and reciprocating to others. Are network members unable to provide more assistance due to lack of skills or knowledge, or have they provided support in the past and are now unwilling or unable to do so? Overwhelming family stressors, such as homelessness, may be present to interfere with the provision of support.
8. How are social support needs prioritized in relation to other presenting problems and needs? What other levels of the environment need further examination?

The guidelines for assessing social support are useful for both worker and client to consider. At a minimum, the worker and client should be able to identify the structure of the network and the various patterns of support or conflict, assess the adequacy of the network for providing support, and pinpoint sources of conflict, identifying a strategy for dealing with at least one of them. The social network map has other uses when administered repeatedly over set time periods; for example, the data can be used to track changes in network structure or functioning over time and thereby be used to monitor and evaluate progress toward goals. As a precursor to establishing network goals, clients may also be asked to visualize an ideal network, as they would like it to be in the future. This visualizing activity leads quite easily to the generation of goals; for example, if the client wanted to have more friends in his or her network, then the worker and client could begin to identify current obstacles and the first steps toward that goal.

The social network map should be considered in conjunction with other key pieces of information about the client and the client's environment. The information gathered from the social network map will often yield clues as to which other aspects of the environment are most important to assess. For example, a young mother living in a housing project may describe no relationships with neighbors due to the dangerous nature of the environment; in this instance further assessment of the physical environment and community/neighborhood setting would be indicated. A recent immigrant to the Unites States may describe a small network that is not able to link the client to vital resources; the worker may decide to use an ecomap or a cultural gram to explore culturally relevant connections with community resources. A parent, among the working poor, may describe overwhelming stressors, such as lack of health care insurance; the worker may opt for narrative assessment techniques to get at the client's experience and/or techniques of power analysis to assess larger social and political forces. As a final example, a parent involved with child protective services may identify a social network limited primarily to kin and formal service agencies; the worker may find it helpful to follow up to determine if insularity, that is, highly negative interactions with kin or formal service systems, is present. In previous work with high-risk families, insularity has been associated with poor maintenance of behavioral changes.

Guidelines for Social Network/Support Goal-Setting

The most obvious use of environmental assessment information lies in the generation of appropriate change goals with the client. Typically goals in social work practice are prioritized in terms of such factors as how immediately the problem must be addressed, how important the issue is to the client, and how likely the intervention is to succeed in relation to other targeted areas for change. In prioritizing environmental assessment information with other sources of information about the client, it may become apparent that an environmental change must take place before other changes can occur; for example, the physical environment may be unsafe or below minimal standards of living, as with a client with no permanent living arrangements, no electricity, or substandard housing. In the same manner, the social environment may be noxious or stress-producing, as with a client living in a crack house, a child physically or sexually abused by another household member, or a "sandwich generation" adult caught between multiple and competing care giving roles. In other instances, a change in the environment may support other targeted changes occurring in the client's life. Examples here would include participation in self-help groups, such as Alcoholics Anonymous or Parents Anonymous, as part of participation in other formal counseling services; participation in a residential program as part of substance abuse treatment for a mother; and involvement in a family support program as part of an early childhood intervention program.

In terms of social support, change goals may be directed toward either structural changes in the social network or functional changes in social network relationships. The choice of goal is dependent on an valuation of the current social network and the changes that are desired by the client in this and other areas. As with goal selection in general, the client's expressed desires direct the choice of goals. Preferences and beliefs about help-seeking influence whether or not mobilization and utilization of social network resources would be considered an option. People hold different attitudes about asking for and receiving help from others, and their ideas are often culturally and socially determined (e.g., a good daughter "should," a grandmother "should"). In extreme cases of "negative network orientation", it is thought to be dangerous and inadvisable to utilize network resources. In such cases, the client may have a network but make little use of it or avoid its use for the purposes of social support. Given client input and agreement, some examples of social support goals and accompanying interventions would include:

- increase or mobilize various types of needed support (e.g., concrete, emotional, or informational, as in enlisting volunteer transportation or respite care services for a family);
- change the composition of an individual or family social network (e.g., helping a person develop new friendships or linking a person with an informal community resource, thus reducing heavy reliance on formal services);
- increase skills in developing and maintaining supportive relationships (e.g., teaching social and interpersonal communication skills or teaching ways of dealing with network members who do not reinforce change efforts);
- improve or enhance the functioning of social network relationships (e.g., helping a family learn to reduce arguments among themselves or developing strategies for a person to use to handle criticism from others).

Most social work settings will require goals to be expressed or written in clear, behaviorally specific language. Dialogue with the client is often necessary to reach clear goal statements. Specific objectives are to be preferred over vague goals, because clearly specified targets for change can help maintain the vision and mobilize action toward the goal. For example, rather than establishing a goal "to increase social support," more behaviorally specific goal statements might be "to have at least one friend or contact in your neighborhood" or "to have one concrete service [such as transportation] provided by a network member, rather than by the social worker." Goal attainment scaling can be a useful way to document progress toward goal completion and to identify the sequence of steps that may need to take place in order to reach the final objective.

The choice of goal should also be based on an assessment of obstacles that prevent the client's benefiting from social support. For some clients, the major obstacle will be lack of or limited social network resources, or network resources that have been "burned out" or otherwise depleted by long periods of care giving. For other clients, the social environment assessment may yield a stress-producing, non-supportive, or antisocial network. Other clients may lack the social skills necessary to form and maintain interpersonal relationships; for these clients, interventions to address skills deficits may need to precede direct linking activities to strengthen social networks. Prejudice, stigma, and discrimination may be factors limiting social networks; for example, the stigma of mental illness has a negative effect on case managers' abilities to mobilize social network resources for persons with chronic mental illness.

Depending on the practice setting, systems barriers, such as paperwork, caseload size, and lack of time may prevent enhancement of social networks. Finally, the success of many social network interventions is highly related to the level of resources in the community and the worker's ability to organize and develop formal and informal community resources. Experience from at least one social support intervention project within the mental health system suggests that the implementation of social network interventions requires a blending of both direct clinical and community organization practice skills. Working toward changes in a client's or family's social support network may also require working toward a change in the service delivery system, the development of new community resources and services, or the forging of new linkages and relationships with informal resources. The next chapter will explore a range of environmental interventions that emerge from social network interventions.

Summary

We end this chapter with a summary table: "Environmental" Assessment Tools and Methods", which contains a wide range of environmental assessment techniques across different levels of client systems and environments. This information is not meant to be exhaustive, but rather to illustrate the range of strategies and techniques that are currently available or being developed to assess various aspects of the environment. The interventions described here also meet many of the characteristics set forth earlier for in the definition of environmental assessment. All of these tools offer a means for the client or groups of clients to learn more about their environment — its strengths as well as its limitations — and its impact on their lives.

Environmental Assessment: Tools and Methods

Perceived Support Network Inventory: A self-report paper and pencil questionnaire of perceived social support; respondents first generate a social network list and then rate each network member in terms of the type of support provided, satisfaction with support provided, and extent to which network member is asked to help and is responsive to requests for help, reciprocity, and conflicts.

Multidimensional Scale of Perceived Social Support (MSPSS): A twelve-item self-report paper-and pencil instrument designed to measure perceived social support from three sources – family, friends, and a significant other; higher scores reflect higher perceived social support.

Sociopolitical Control Scale (SPCS): A seventeen-item self-report paper-and-pencil instrument to measure sociopolitical control as distinguished from other types of perceived control; the scale can be used to measure a sense of empowerment in clients; higher scores indicate higher perceived leadership competence and policy control.

Family Support Scale: An eighteen-item self-report paper-and-pencil measure of the perceived helpfulness of various sources of support to families raising young children.

Narrative Techniques: Narrative methods, such as case studies, self-characterizations, and repertory grids, take into account life stories, narratives, and myths, and the meanings that humans assign to their individual and collective experiences .

Ethnographic Interviewing: Interview techniques that seek to help workers understand cultural differences and value clients as cultural guides, explaining in their own words the meanings they attach to problems, relationships, and the help-seeking process, among others.

Participant Observation: Method of assessment whereby the worker observes the client in everyday life and participates in the client's daily routines as unobtrusively as possible; especially useful when the view of the "insider" is needed or difficult to obtain; can be used with individual clients (e.g., observing a student in class), families (e.g., having dinner with a family), groups (e.g., attending a tenant group's meeting), and at the neighborhood level (e.g., participating in daily life in a housing project).

Organizational History of Neighborhood: Guidelines for organizing information about a neighborhood, considering recent history, divisive factors, organizational ties, identifiable neighborhood leaders, and individuals roles in community activities.

Physical Environment

Environmental Assessment Index (EAI): A forty-four-item instrument to assess the educational/developmental quality of children's home environments via information gathered through a home interview with the caretaker and direct observation by the worker.

Family Access to Basic Resources: An assessment tool completed by both the worker and family to determine the extent of stress and stress pileup due to inadequate or unstable basic family resources; includes monthly expenses, current resources, potential family resources, and resource stability for the following resources: wages, child support, income transfers, housing, food, clothing, personal care and recreation, health care, education, family and developmental services, and transportation.

Family Resource Scale: A thirty-one-item self-report paper-and-pencil questionnaire to measure the adequacy of different resources in households with young children; each item is rated on a five-point scale ranging from Not at All Adequate to Almost Always Adequate.

Physical Description of Neighborhood: Guidelines for describing the physical features of a neighborhood, including mapping of each block, types of automobiles, physical evidence of political activity, general upkeep, physical boundaries, number and variety of behavior settings, and extent of similarities and differences in the neighborhood.

Social/Interactional Environment

Social Network Map: Visual means to depict size and composition of social network, types of social support perceived to be available, and quality of relationships within a social network.

Ecomap: Visual means of placing an individual or family within the social context, particularly the organizations and factors impinging on their lives.

Child's Ecomap: Ecomap designed for completion by child entering placement; covers areas such as homes the child has lived in, friends, siblings, school, dreams, worries, and feelings.

Community Interaction Checklist: A self observational listing of types of interactions occurring in the past twenty-four hours (or other) time period with members of the social network outside immediate household members; the nature and positive or negative rating of each interaction is recorded by the client.

Inventory of Social Support: A paper-and-pencil matrix of personal social network in terms both of source and type of support; respondents indicate whom they would go to or receive help from for each of twelve different types of child-rearing supports.

Sociograms: Graphically depicts patterns of affiliations and relationships between group members by using symbols for people and interactions; useful in identifying and assessing subdivisions and their impact on a group.

Socio environmental Context of Group: A visual representation of a group's situation and the environmental forces that may be critical to the group's functioning; exchanges that represent positive support for or opposition to the group from a variety of sources are depicted, including the larger agency structure, the families of the members and worker/facilitator, and other systems specific to the group, such as funding sources

Nomothetic Ecomapping: Visual display of common supports and obstacles from a variety of systems for a particular neighborhood-level concern (e.g., school failure); results in a generic ecomap that may be used to inform choice points for intervention.

Institutional/Organizational

Nurturing/Sustaining Environment: Graphic representation of both the supports and problems a person experiences in the social environment; the positive, negative, or neutral effects of both the nurturing (immediate) environment and the sustaining environment (wider community and broader society) are identified; information can be generated with the client and leads to understanding of areas that need to be changed and areas of strength that might be resources for achieving change.

Family Empowerment Scale: A thirty four-item self-report paper-and-pencil instrument to measure empowerment of a parent or caregiver of a child with an emotional disability; the level and expression (through attitude, knowledge, and behaviors) of empowerment are assessed.

Context Diagram of Environmental Transactions: A visual representation of the transactions and exchanges occurring between an agency and other systems in its environment (e.g., clients, community agencies, schools, other service systems, etc.).

Force Field Analysis: A technique to organize information about the forces that may affect the success of a change effort: the steps include (a) specifying the desired change (usually at the organizational or community level), (b) identifying the driving and restraining forces that will affect whether or not the objective is met, (c) assessing the strength of each driving and restraining force in terms of potency, consistency, and amenability, (d) identifying the actors for each driving and restraining force and rating the likelihood that the actor will become involved, and (e) selecting a strategy for change based on the analysis of the various forces working for or against the change effort and the various actors that may be able to help.

Mapping Community Capacity: A format for assessing the assets and capacities existing within communities;

three levels of assets are examined, those assets and capacities located inside the neighborhood and largely under neighborhood control (e.g., individual assets of residents and local resident-controlled associations and organizations), assets located within the community but largely controlled by outsiders (e.g., private and nonprofit organizations, physical resources, public institutions and services), and resources originating outside the neighborhood and controlled by outsiders (e.g., welfare expenditures, public capital improvement expenditures, and public information).

Inventory of Groups and Organizations: A booklet-length guide to finding out about the organizations and groups people belong to a city neighborhood; three methods of gathering information are described: using newspapers, directories, and other printed sources; talking to people at local institutions, such as parks and churches; and conducting a phone survey of a sample of local residents.

Frameworks for Analyzing a Community or Neighborhood: Community assessment leads to organized information about strengths and limitations of the community, the manner in which the community solves problems, and the capacity for change. Most organizing frameworks consider such factors as physical setting, history and demography, economic and political system, and educational, health, and welfare systems.

Social/Political/Cultural

Culturalgram: An ecomaplike assessment tool designed to assess the impact of several different aspects of culture on the family or individual, including reasons for immigration, length of time in the community, legal or undocumented status, language spoken at home and in the community, contact with cultural institutions, health beliefs, holidays and special events, impact of crisis events, and values about family, education, and work; also enables the worker to individualize for ethnically similar clients beyond cultural generalizations.

Power Analysis: A critical component of empowerment-based practice, a power analysis is conducted with a client in two phases: first analyzing how conditions of powerlessness are affecting the client's current situation and second identifying sources of potential power available to or within the client.

Cultural genogram: Technique described to add cultural information to a genogram, including dimensions such as groups represented in culture of origin and intercultural marriages.

Framework for Conceptualizing Community: Format for organizing information collected about a community or neighborhood; the factors to be considered include understanding characteristics of target population members, identifying community boundaries, profiling social problems, understanding dominant values, identifying formal and covert mechanisms of oppression, identifying evidence of discrimination, recognizing locations of power, determining resource availability, and identifying patterns of resource delivery and service delivery.

6

WATERSHED MANAGEMENT

Despite his advances in science and technology, his sophistication and his many accomplishments, man owes his existence to a mere six inch layer of topsoil. This topsoil is, however, being lost at the alarming rate of 12 million tons per annum due to loss of vegetal cover, overgrazing and agents of erosion like wind and water. Since nature has thus richly endowed' the earth with soil and water, it is up to us to preserve both, especially so in those areas which receive erratic rainfall and occasional storms - the dryland areas, still depending only on the rainfall for moisture. India has a land mass of 329 million hectares, out of which 224 million hectares constitute the dryland areas. They are undulating or rolling lands and depend only on rainfall for any biotic growth. Since the past 2 decades, productivity of dry land crops like jowar, tur and groundnut has remained low and stagnant.

Production over time

Dryland crops	*Average yields*	*(kgs per hectare)*
	1969-70	*1987-88*
Jowar	530	655
Tur	687	722
Groundnut	712	800

Source: Government of India, Status Report on Agriculture, 1989.

This, in spite of the fact that India's annual rainfall is next only to that of South America and much more than anywhere else in the world.

Reasons of Low Productivity in Dryland Areas

Though India has an annual average precipitation of 1150 mm and drains a total volume of 400 million hectare metres, hardly 100 million hectare metres is retained in the soil. The remaining water runs off, carrying with it fertile soil.

Dryland areas are characterised by undulating terrains with varying soil depths. When it rains, the rain water accumulates in the top most rocky areas and gushes down, damaging the lands below. The lands below, termed as common property resources, are neglected. Even small bushes are hacked by the neighboring populace to meet the increasing fuel demand. Grazing livestock peel off the grass cover and trample the soil with their hooves. Population pressure has forced farmers to bring some of these lands also under the plough. With no cover and canopy left, the slightest splash of raindrops dislodges the soil and the runoff carries it down the slopes. Over a period of time, the base material gets exposed. The runoff further enters the fields down the line causing rills,- stagnating against the bunds and damaging crops. Incapable of holding the force of the runoff, some of these bunds get breached.

The water with the soil cascades, breaching more bunds downstream, leading to the formation of ravine's in fragile land.

Hence when it rains, rainwater moves with great velocity carrying soil. The nallas in the area overflow with gushing, muddy water. With very little infiltration and 'loss in both soil and water, productivity is lower. Farmers are left with no surplus to compensate this loss of soil and nutrients and are caught in a vicious circle. Farmers in dryland areas with low rainfall of 500 mm or less, have suffered most in this process, for their land has slowly lost productivity in all types of biomass production.

This is the fate of the lands contributing most of our production of millets, pulses and oil seeds.

Solution of Problem

Gone are the days when man blamed the natural occurrences of weather and rain. Modern methods prove that even in the scanty rainfall areas, productivity can be bettered. A resource management that conserves and uses rainfall and the fertility of the soil more efficiently and a new production, system which maintains productivity and assures a dependable harvest, have now been developed. Watershed-based farming appears to be a development alternative that would provide greater stability to rain fed agriculture.

Definition of Watershed

A watershed is an area of land from which surface water drains into a single outlet like a stream or a river or a lake. The rain water flows from the ridges, along the slopes, into the nallas and finally collects in the village ponds. This entire area with one common drainage is called a watershed.

In general, the watershed area for a small stream may be a few hectares, while for a large river the watershed may be many square kilometres. For proper planning and execution, the size of a watershed area should ideally be 1,000 to 2,500 hectares. If the area is within this range, it will be possible to prepare a well balanced plan and to implement it in a period of 2 to 3 years.

Objectives of Watershed Management

Watershed management is an attempt to halt land degradation and a holistic process for getting maximum production out of land. Watershed management implies rational utilisation of land and water resources for optimum and sustained production, with the minimum of hazard to natural resources. With a bit of acumen and proper judgement, the rain water can be made to move down the slopes slowly, ensuring optimum infiltration and percolation, automatically solving the problem of soil erosion. Reducing the impact of the rain on the soil, checking its speed at various intervals, taking up all operations on the contour and diverting the excess to prevent the, pressure, are some of the procedures one could adopt; a process that starts at the highest point of the watershed and ends right down in the lower parts.

In general, watershed management consists of proper management on the hilltops, hill slopes, foothills and plains with appropriate practices. Hilltops should be kept under forest, and hill slopes up to 25% should be kept under plantations (either horticulture or afforestation). Foothills should always have cover crops to avoid formation of gullies and siltation of plain lands at the foothills. Proper diversion drains for the runoff water from upper catchments in a regulated manner in the foot hills will increase the water availability for irrigation in the plains, by means of wells, tube wells, tails, etc.

Below the foothills., lands should be treated with soil and water conservation measures, either biological or mechanical to avoid sheet erosion and at the same time to preserve the top soil for maintaining the yield of agricultural crops. Finally, plain lands, banks of small streams and rivulets are usually kept under agricultural or horticultural crops with irrigation facilities wherever possible. Cropping patterns are adopted whose growing period is most suited to the amount and periodicity of rainfall.

Steps in Watershed Planning

- Preparation of base maps for carrying out surveys.
- Reconnaissance survey of the watershed for overall development.
- Assessing rainfall characteristics.
- Preparation of soil maps and classification of lands for different uses according to capability classification for agriculture, forestry, pasture, horticulture, etc.
- Preparation of inventory of existing land uses and farm sizes.
- Appraisal of agricultural production patterns and potentials, present and potential markets and possible group action arrangements.
- Carrying out topographic and hydrologic surveys for engineering works.
- Geo-hydrological survey to delineate areas suitable for groundwater development.
- Formulation of an integrated time-bound plan for land and moisture conservation, ground water recharge, development

of productive afforestation, agriculture production, grass lands and horticulture.

- Assigning of priorities for implementation of the project.
- Assessing social costs and benefits.

Land Capability Classification

Watershed management must take into account land capability classification, since a typical watershed may include lands suitable for cropping, trees, pastures, etc. The land classification is as follows:

Class I:

This type of land is nearly level, with minimum erosion, the soils are deep, well drained, with adequate water holding capacity. These soils need ordinary management practices to maintain productivity. Such practices may include the use of one or more of the following fertilisers, lime cover and green manure crops, conservation of crop residues aAd crop rotation. They are fit for agriculture, pasture, forests, etc.

Class II:

These soils have some limitations. They are subject to slight erosion, as they are slightly sloping lands, with less soil depth, occasional drainage problems that can be easily corrected, and they are sometimes saline. They may need one or more of the following practices terracing, strip-cropping, contour cultivation, excess water drainage, crop rotation, use of green manure crops, stubble mulching, the use of fertilisers, manure and lime.

These soils may be used for growing cultivated crops, raising pastures, forests, etc.

Class III:

The factors that limit plant growth are more severe than in class II. These factors could be moderately sloping land, land susceptible to water or wind erosion, land more susceptible to drainage problems, land with water-logging, shallow soil depth, low moisture holding capacity and moderate salinity.

Class IV:

The limitations to plant growth is more severe here than in Class 111 soils. Very careful management is required and the conservation practices are more difficult to apply and maintain. Some of the factors affecting these lands are steep slopes, severe susceptibility to water and wind erosion, shallow soils, low moisture holding capacity, severe water-logging and severe salinity. These soils can be used for crops, pastures, forests and wildlife food, etc.

Class V:

Cultivation is not feasible because of one or more limitations, such as over flow, stoniness, wetness or severe climatic conditions. Examples of this class are soils of low lands subject to frequent over flows which prevent the normal production of cultivated crops, or rocky or stony terrain. These soils are suitable for raising cultivated crops but are not suitable for perennial vegetation.

Class VI:

These soils have very severe limitations that make them generally unsuitable for cultivation and limit their use largely to pastures or forests or wildlife food and cover. Some of the limitations which cannot be corrected are: steep slope, very severe erosion hazard, shallow rooting zone, excessive wetness, low moisture capacity and severe salinity.

Class VII:

The physical condition of the soil is such that it is not practical to adopt pasture improvements and water control practices.

Class VIII:

Bad lands, rocky out crops, sandy beaches, marshes, deserts, mine tailings and other nearly barren lands may be included in this class. It may be necessary to give protection and management for plant growth to soils and land forms in class VIII in order to protect other more valuable soils, to control water, or for wildlife or for aesthetic reasons.

Water Management Activities

In situ conservation practices. The best opportunities for moisture conservation lie in the control of the portion of precipitation which is normal runoff. Retaining surface water on the land surface for a longer time, increases the amount of water entering the soil. Effective methods are the following:

- Contour bunding / graded bunding
- Check dams and gully control structures
- Land levelling / land smoothening
- Bench terracing
- Farm ponds
- Percolation ponds
- Waterways
- Diversion drains

Water conservation starts in the topmost rocky areas where the runoff water needs to be drained out by forming diversion drains. This protects the steep slopes below which are suited only for growing fodder. While prevention of grazing in these areas ensures optimum regeneration, luxuriant fodder cut and fed to the animals ensures better nutrition and health.

Even on similar private lands, it is always advisable to grow fodder or fibrous crops like agave.

On the lands just below, which are slopes with moderate soil depths, if belonging to the public, V-shaped trenches can be formed which act as speed breakers to the runoff.

Trenches around the planting can be made with necessary spacing according to the intensity of the rainfall. Trenches 3 to 4 metres apart should be dug, depending on the degree of the slope. It should have a width of 30 cms at the bottom and a width of 50 cms at the top with a depth of 30 to 40 cms., depending on the available soil depth. The soil dug from the trench should be heaped to form a mound on the slope of the hill side where seeds of hardy species can be sown. Mixed species can be planted in these trenches to have a staggered harvest, ensuring a permanent canopy. Fodder can also be grown in between the trenches. Nurturing and protecting these areas can be taken up on a community basis. Strict fire control measures and prevention of cattle grazing will enable good fodder production on these lands, which can be cut every year for stall feeding of cattle. The produce can be shared by the entire community.

If the lands are privately owned, the farmers can resort to farm forestry or horticulture. Initially the farmer may feel insecure. Hence, agricultural activities may be continued between the rows of trees.

As the trees grow, though agricultural income would drop, the trees would start yielding. This would definitely prove lucrative to the farmer. Thus, if immediate switch over is not possible, strip cropping can be resorted to, to a seasonal cultivation between the wide rows of trees. The runoff- from the agricultural strips is checked as it reaches the row of trees.

For optimum conservation in these lands, contour bunding would be the answer.

Contour/Graded Bunding

A contour is a line joining points of the same altitude on a sloping land. Bunds are constructed along these lines. Contour bunds are suggested for light soils with low rainfall and graded bunds for all black and red soils with medium to heavy rainfall with necessary outlets for runoff disposal. Wherever the black soil type exists, graded bunding is adopted to drain off excess water to the nearest water streamlet because percolation is slow in black soil and therefore more runoff is expected than in red soil. These types of bunding are adopted upto slopes of 3% - 5%. According to the runoff, disposal outlets should be provided along the bunds to reduce the breaching of the contour bunds.

Bench terracing is adopted on hill slopes to raise commercial or food crops, instead of a plantation programme. Before taking up bench terracing, it is necessary to confirm whether enough soil depth is available for making terraces. The terraces can have either an inward or outward slope. The verticals are not available for cultivation but suitable for grass or stone pitching to keep the vertical intact. The width of the terrace depends on the slope of the land. A minimum of 1 1/2 to 2 metres level ground is required for cultivation., According to the rainfall, drainage of the verticals has to be established at 30 to 40 metres intervals for safe disposal of runoff water. In high intensity rainfall areas, the verticals should be provided with rivetments to avoid erosion.

A permanent canopy of bushes, trees, or fodder on the slopes, whether private or public property, can step up productivity and go a long way in controlling erosion. The treatments mentioned above can ensure insitu rainwater conservation during normal rains, and, the water flows out slowly and safely, without sedimentation. But during heavy rainfall the excess is prevented from entering into the agricultural lands below by constructing diversion drains.

Diversion Drains

Diversion drains should be constructed at foothills or at the junction of the cultivated land to avoid silting up of the cultivated lands due to runoff water flowing down the hill and across the cultivated lands. In addition to this, diversion bunds have to be constructed at the junctions of roads and cultivated lands to dispose the runoff water to the nearest waterway, nalla or streamlet along the diversion bunds. Quick growing trees have to be planted along the bunds to strengthen them so that there is no breaching of the bunds during heavy rainfall. The bunds may be planted with brushwood or species like casuarina, eucalyptus, agave etc.

Land Levelling/Land Smoothening

For cultivation of agricultural crops generally in dry lands, it is levelled with a meagre 112 % slope to avoid accumulation of water during rainfalls and at. the same time drainage disposal is provided. Land levelling or smoothening is done either by bullock. drawn wooden frame levellers, or by tractors. In case of undulating land with rolling topography, land levelling is done using bulldozers. During land levelling operations, care should be taken that the top soil is not lost, thereby reducing the yield from these lands. Mixed crop or intercrop cultivation can be adopted to get maximum benefit even during drought conditions.

Mulching

Use of green manure and mulching ensures better retention of water and also increases the organic content in the soil.

Contour Ploughing

Ploughings are carried out in contour. At every 3 metres, deep contour furrows, are opened and a small section bund is formed every 10 metres to check the flow of rain water. These types of ridges and furrows formed along the contour at intervals of 3 to 6 metres help to form plots for moisture conservation during rainfall and to reduce loss of soil. With ploughing and sowing done on a contour, the rain water gets distributed uniformly, leaving no scope for rills and gullies. In widely spaced crops like tur, deep furrows are opened for every row.

These practices ensure moisture conservation during normal rains. But during heavy rains, after the soil is saturated, the water starts moving faster. To arrest this momentum, bunds are constructed which are stabilised with a vegetative cover.

Recent developments suggest the supplementing of soil bunds with vegetative barriers. For this purpose, a deep rooted grass, like khus/vetiver with a tendency to grow thick is ideal. Here, water does not stagnate. Instead, the silt is deposited on the upstream, and the filtered water oozes out all along the barriers. Under experimental conditions, vegetative hedges with different types of grasses have proved very successful.

The excess water along the bunds is channelised through grass-lined outlets into waterways.

In watershed development activities, waterways are used for the safe disposal of runoff water near foothills areas and at the junctions of agricultural and non-agricultural lands without silting up the cultivated lands. The waterways are connected to a drainage system. On plain lands, foothills or hill slopes, the existing waterways are already in an eroded or silted condition due to natural reasons like high intensity rainfall or concentration, of water in one place due to the bed slope in the gullies and valleys, etc. The waterways have to be stabilised either with grass or with stone rivetment or with stone pitching, according to the construction material locally available.

In addition to this, necessary checks should also be constructed along the waterways for storage of runoff water and to reduce the bed erosion of the waterways.

Through these waterways, water is collected in the farm ponds.

Harvesting runoff into small ponds and recycling this water to the donor area is a recent approach. The size of the pond needs to be 250 to 300 metres per hectare of catchment area. In red soil areas, the percolation losses are high. To reduce such a loss, a mixture of soil and cement (8:1), coal tar (bitumen), and thick plastic sheets are used to avoid seepage. The water collected in the farm ponds can be used to give one to two life-saving irrigations to the rain fed crops. Effective irrigation can be carried out by means of portable pump sets. Farm ponds should be provided with stone bed pitching and an outlet and a silt trap to avoid silting up of the pond.

There will still be runoff in the fields even after erosion prevention measures have been taken and farm ponds constructed. The excess water may be collected by constructing percolation ponds at appropriate places.

Percolation Ponds

Percolation ponds are to be located in such areas so that they may not interfere with existing irrigation tanks. Usually, the location should be in poramboke lands. They should not be located in heavy soil or in impervious strata. The size of the percolation pond varies according to the runoff. It will usually be located upstream of the cultivated lands. Deep ponds are preferred because evaporation is minimum as also the area of land submerged. The area selected for the percolation pond should be treated. Earthern embankments should be formed from the available soil nearby. A disposal weir and an overflow waterway should be provided to avoid breaching of the embankments.

In addition to runoff water collection, the percolation ponds are helpful in recharging the ground water in the area and stabilising irrigation. It is estimated that the pond may be effective up to a distance of 1000 metres, depending on the strata in that area.

Despite all the care, if there is still excess water, it is led into the nallas. Through vegetative cover, the nallas courses are stabilised. Gully checks with vegetative or brushwood dams help control the speed of flow of water in the nalla.

Check Dams and Gully Control Structures

A gully is formed due to improper drainage disposal when water accumulates in a single place either due to natural reasons or due to artificial conditions. Before the formation of gullies, rills are formed on gentle sloping land depending on, the runoff. They deepen quickly and form small gullies. To prevent the formation of deep gullies and to fill up the drainage course, gully control structures should be provided at appropriate places with locally available materials. Check dams could be Constructed across small streams and gullies which would act as mini percolation ponds having a catchment area of less than 20 hectares. The length and breadth of the check dam will vary according to the width and depth of the gully or nalla- Brushwood or stone checks can be constructed in rows at certain vertical intervals according to the bed slope of the gully. Nalla bunding is also a gully check on a larger scale where the runoff water is also high.

The control measure requires a permanent masonry structure like a weir across the nalla and the strengthening of the embankments on either side of the nalla. Accumulation of too much water in the upstream side of the gully should be diverted by means of diversion drains or grass covered waterways to the nearest stream or overflow weir. To avoid erosion in the waterway, whatever there is a difference in the elevation of more than 2 metres, drop structures should be constructed either with a vegetative structure or with bed pitching, according to the discharge in the overflow portion of the waterway.

If there are a number of shallow or medium gullies in the waterway, stone check structures should be constructed to protect it from further erosion and to facilitate growth of vegetation. It is also possible to establish in situ stone structures at 100 to 200 metres horizontal intervals along the gullies in the watershed.

The upper catchment of the gully should be covered with vegetative crops or grass to reduce the runoff. Gully heads can be checked with wooden poles or stones to avoid entering of the gully into the upstream areas. Check Dams may be constructed according to the bed slope, either with an overflow weir/stone filled checkdams with oven wire of small size.

Constructed in the steeper sections, are ravine control structures. So water flows slowly, safely and without sedimentation even in the nallas. This clear water is impounded behind nalla bunds which are either earthern or masonry.

Thus, due to all the treatments from the ridge to the valley, along all the arable and non-arable lands, when it rains water moves slowly, ensuring higher infiltration and, even better, underground recharge. With this kind of soil conservation, enrichment and better moisture retention, farmers can get a better yield even with their present farming and cropping practices.

When there is a prolonged dry spell, the water stored by the various methods already mentioned may be used to give supplemental irrigation.

Sprinkler and Drip Irrigation

To economise the use of water, sprinkler irrigation can be used for crops such as millet, oilseed and pulses, and drip irrigation can be used for horticultural crops. Sprinkler units are hired out by government agencies or co-operatives.

Sprinkler and drip irrigation Old method of drip irrigation Modern method of drip irrogation

Use of Ground Water

Watershed schemes offer an increase in ground water potential. To recharge ground water flow in the watershed, small subsurface dams may be constructed. The available ground water in the watershed should be pumped and better water management practices should be observed.

Dryland Crops

With moisture available in the top and bottom layers of the soil, a mix of shallow and deep rooted crops or even double crops can be planted. With good biomass planning, farmers can obtain fodder, fuel and fruits from the upper portions of lands, cereals and pulses from the middle, and commercial crops like vegetables, fruits and flowers from the lower areas where water from the farm ponds can be used for protective irrigation.

For the better utilisation of soil moisture, farmers can plan a combination of third rows of tur in crops of ragi, groundnut or soyabean. Alternatively, these crops can be rotated with tur. Tur fixes nitrogen and produces a lot of leaf fall. Its stock is also a fuel. Maize and tur is another good combination. In a base crop of sorghum, green gram, black gram or cowpeas can be raised as inter crops. In bajra, cowpeas or cluster beans is the best combination crop. With cotton, black gram or coriander is suitable for intercropping. The contoured furrows opened between the rows of tur ensure better conservation of moisture. Moisture and nutrients from the upper regions of Soil is absorbed by shallow rooted jowar or groundnut, while deep rooted tur absorbs them from the lower portions. When the moonsoon fails, the jowar or the groundnut wilts due to evaporation of surface moisture, but tur is a survivor and uses the still available moisture in the deeper layers.

This takes care of the farmers risk to a certain extent. Besides, with normal rains, when both crops mature, the farmer is assured of cash at different points of time. Relay cropping or sequential cropping should be adopted, depending on the conditions in the watershed area. The cropping intensity will vary according to the soil type and the amount of rainfall.

Weed Control

Weeds compete with crops for water and nutrients, thus bringing down the yield. Therefore, timely and effective weed control measures will increase production of dry land crops by 25 to 50% over unweeded crops. Sorghum should be kept weed-free from 21 days after germination upto 35 days to obtain maximum yield. For cotton, the weed-free Period must be between 30 to 50 days after germination.

Optimum Plant Plantation

Optimum plant density is ensured by farmers for optimum assured yields. For example, for sorghum there should be 148,000 plants per hectare for maximum yield. Seed drill sowing with compaction is the best way to maintain optimum plant population and achieve higher yield.

Low Cost Fertilisation

Dry land crops receiving a balanced supply of nutrients through fertilisers have been found to be more tolerant to drought than crops that have not been fertilised. Nutrients supplied through t combination of organic (farmyard manure), inorganic (chemical fertilisers), and bio-fertilisers (rhizobium) help to improve the yield of rain-fed crops.

Analysis of Rainfall Patterns

Analysis of long-term rainfall data to assess the quantity, distribution and probability of occurance is a very important technique. Annual, seasonal, monthly and weekly rainfall data are analysed to determine the distribution pattern, the beginning and end of the rainy season, the timing and duration of wet and dry spells and the probability of rainfall. The probable rainfall with regard to water holding capacity of the soil, and evaporation demand, can be used to develop the soil-moisture availability pattern during the growing season, and a suitable cropping pattern can thus be suggested.

Dry Seeding before the Monsoons

Dry seeding before the onset of the monsoons, especially in black soils is helpful in making better use of the initial rains. The sowing of sorghum and cotton 2 and 3 weeks respectively before the monsoon season, has been found to give better yield. Sowing at an optimum depth is important. Sorghum and cotton should be sown at a depth of 5 cms and bajra, black gram and other grain crops at a depth of 3 cms. Pre-hardened seeds develop an efficient root system, improve drought tolerance and have vigorous initial growth.

Use of Improved Crop Varieties

The combination of genes for dwarf plant stature and resistance to lodging, with genes responsive to applied nutrients and drought tolerance, is proving successful with sorghum, cotton and millet. Improved varieties and hybrids significantly out-yield local varieties in most of the rainfed areas.

Agricultural Implements and Machinery

Ploughing the field after the harvesting of the crop, land levelling and shaping formation of field drains and community drains, broadbed furrow sowing, application of fertiliser, weeding, etc., require tractors and improved bullock-drawn and power-operated implements that will permit quick and efficient completion of the various operations. These implements should be made available in the watershed area through government agencies or community organisations. Customers Hiring Units may be formed for the purpose.

Plantation Programmes in Community Lands

Community lands which are owned by panchayats or by the Government are kept fallow. These lands have rocky outcrops and no adequate soil depth due to the eroded condition of the soil. No proper soil or water conservation management practices or treatment are carried out on these lands. They should be brought under social forestry programmes or under pasture development according to the type of land and its suitability. The lands which are slightly away from the villages may be brought under pasture development by means of water harvesting from the upper catchment areas. But the planted area has to be strictly protected from grazing for at least five years to ensure the survival of the planted seedlings.

In case of plantations in wastelands, the available annual fodder and grass can be increased if it is treated by means of trenches having suitable grass species namely, Stylosanthes hamata, Dicanthium annulatum, etc. In social forestry plantation programmes, leguminous fodder trees, such as subabul, and other quick growing trees can be planted, with necessary protective measures taken for the first five years. If possible, pot irrigation should be taken up in summer. To avoid over grazing, stall feeding of cattle right from the second year should be resorted to. In addition to this, incase of badly eroded and gullied lands, along the contour agave plantation can be taken up with proper spacing and a proper drainage system with diversion drains, gully plugs, check dams, nalla bunds, etc. In case of heavily eroded lands, it is necessary to take constructed protection methods like fences to prevent cattle grazing. The best method to develop community land is to lease it either for pasture or for plantation programmes with inter-cultivation wherever feasible.

Avenue Plantation

Along the roads in the districts, villages and urban areas, avenue plantation raising can be taken up. Two rows of trees should be planted on either side of the road, the first row to improve the aesthetics and to produce, in course of time, several items of minor forest produce for collection and use by local villagers. The second row of trees would produce fuelwood and fodder from mostly fast growing hardy species . Plantation programmes should be taken up with necessary techniques like afforestation with pits, soil and moisture conservation programme etc. Proper planning should be done for future widening of roads when avenue plantation is planned.

Development of Wind Breaks and Shelter Belts

It is necessary to have wind breaks in areas where the land is silted up due to wind erosion. A wind break consists of five rows of species planted at an interval of 500 metres. Wind breaks act as a barrier against the wind speed and make the land in between the wind breaks suitable for cultivation. Wind breaks are best adopted in areas where there are sand dunes, to avoid the moving of sand dunes from one place to another. Shelter belts are nothing but wind breaks planted around the periphery of horticultural plantation areas like mango to avoid high wind speed during the flowering season and to prevent the dropping of fruits. The species for a shelter break is selected in such a way that it would not affect the horticulture crop by infecting them with pests, etc.

Community Participation

For a successful watershed development, team work is very essential. A task of this nature requires community efforts. It inculcates a sense of participation. With the responsibility shared and the drive towards a common goal, the fruits of labour can be enjoyed by all. Farmers owning land or having access to public land can seek the assistance of a watershed team, voluntary agencies and banks. For optimum results plant protection seed protection and processing, utilisation of ground water recharge, etc., have to be taken up on a group basis in the watershed. Organisation of inputs like improved seeds, fertilisers, etc., organising bio-mass planning, and ensuring timely operation of critical elements in the watershed, can best be taken up on a community basis. Such a group can also act as an intermediary between banks and farmers.

Productivity in watershed

Dryland crops	*Average yields within Watershed*	*Average yields outside Watershed*
Jowar	1390	055
Tur	1200	722
Groundnut	2000	800

Source: Government of India, Status Report on Agriculture, 1989.

"Even a marginal farmer need not despair. In a. watershed, he can sustain himself and become viable by adopting the agri-horti-model. Out of a total farm area of 0.6 hectares, he can utilise 0.4 hectares for growing crops, and in a year of early rains, he can even grow a cash crop like sesamum. In the remaining 0.2 hectares, he can plant trees and grow some crops in between".

Total area of farn	0.6 hectares
Ragi + Tur	0.4 hectares
Mango plantation + phased reduction of Tur andGroundnut	0.2 hectares

Cost Benerit per year (in Rupees)

Year	*Cost*	*Gross come*	*Net income*
1st year	2910	5655	2745
2-5th years	2470	6095	3625
6-10th years	2406	8041	5635
11-20th years	2406	9049	6447
21st onwards	2252	11830	9578

Source: National Bank for Agriculture and Rural Development.

Participation of Financial Institutions

Bankers participation is very crucial. The bank can draw the resource plans for every farmer in the watershed for optimum bio-mass production and also contingency crop plans to overcome weather uncertainties. The bank is in a unique position to dovetail technology with socioeconomic factors and also the constraints of the individual farmers in the area.

Thus in a compact area, development of a watershed of 1000 to 2500 hectares can be implemented with a multi-disciplinary approach. There is an urgent need to develop an effective system of watershed management. This would be an important step towards the essential improvement of the living conditions of the farmers who subsist on rain fed agriculture.

In conclusion, watershed development, if managed efficiently, can result in an increase in the output of fruits, cereals, pulses, oilseeds and even straw. This is a very rewarding experience for the farmer. But this is just the beginning and with proper planning for the next season, he can reap a richer harvest. So the farmer has learnt to make wise use of the gifts of nature, namely water and soil. Thus, a successful watershed development programme, by conserving water and increasing the fertility of soil, gives greater stability to agriculture and ensures a bountiful harvest.

7

WATER RESOURCES MANAGEMENT

Introduction

No life on earth can exist without water. Scientists estimate that there is over one billion cubic kilometres of water on this earth, which covers nearly three fourths of the earth's surface in the form of oceans, rivers, lakes, snow, glaciers and groundwater. Though this seems an inordinately huge amount, in actual fact, less than one percent is fresh and usable found in lakes, ponds, rivers and groundwater. Of the remaining, 97% is found in oceans and 2% is locked up in glaciers and ice-caps. Only 1% is available for use.

Water for Life

Water is the basic necessity of life, not only for human beings, but also for plants and animals. Life began in water and it is a basic component of every living cell. Water accounts for 65% of our body weight. If we lost even 12% of it, we would die. About 83% of our blood is water. It helps digest our food, take in oxygen, transport body wastes and control body temperature. We need water in almost every domestic activity, from cooking and washing to bathing and sanitation.

Water for Climate

Water, to a large extent, determines climate. Places near large water bodies are cooler because the water bodies act as large sinks for heat, thus moderating the climate of the area. Regions near water bodies generally have milder winters and cooler summers. Water has an even more basic role in climate control through the water cycle. The evaporation of water requires huge amounts of energy, which comes from the sun. When the water vapour falls back to earth as rain, this energy is released. Thus water acts as an energy transfer and storage medium for the climate system.

Water for Agriculture and Industry

Food cannot be produced without water. Vegetables are 80-90% water and milk about 87%. Agriculture is the major consumer of water in India, accounting for nearly 93% of the total water. Almost all industrial processes need water which is needed as a solvent, as a medium, as a coolant, as a cleansing agent, etc. Water plays a very important role in disposing of waste, be it domestic sewage or industrial effluents.

Water for Power Generation

Almost all types of power generation require water - from hydel power, where falling water turns turbines to produce power, to nuclear reactors which need huge amounts of water as a coolant. In thermal power generation, next to the fuel, water is the most important resource. For the production of 1 kilowatt of electricity, thermal plants use 140 litres and nuclear plants use 205 litres of water.

The Water Cycle

From the beginning of time, water has been constantly and continuously in motion. Little has been added or lost over the years. The same molecules have been transferred time and time again from the oceans, rivers and lakes into the atmosphere by evaporation and through condensation, back to the earth as precipitation, which finds its way into seas, rivers and the ground. This is known as the water cycle and it is estimated that at any given time, 5 litres out of every 100,000 litres is in motion.

Water use in India

In India, water is used mainly for agriculture, livestock and industries. An estimate of the total water that will be consumed in India by 2000 A.D. is given below:

Sources of Fresh Water in India

Rivers: The amount of water available for use in India is estimated as 1900 billion cubic metres per year. About 86% of this is the surface run off found in streams, rivers, lakes and ponds. In fact, these are the major sources of water in our country.

Groundwater: India is estimated to have 3,700 mhm. of groundwater, almost 10 times the annual rainfall. Only 10% of the water is being made use of at present, but with lakhs of tube wells being sunk every year to meet the growing water shortage, the water table is declining rapidly.

Tanks: Tanks are small reservoirs built by constructing earthenware dams. They have been in existence in India since ancient times when tanks were built to store rainfall. But in both British and independent India, these tanks have been sadly neglected. As a result, where tanks irrigated half the cropped area a hundred years ago, today they irrigate hardly 10% of it. In Tamilnadu, man-made earthen reservoirs are known as yeris. There are over 40,000 yeris in Tamilnadu. Their place has been taken by big dams which span the major rivers. While the Indian farmer has benefited from these huge hydel projects, and while hydro electric power is a major source of energy in India, the dams themselves have many detrimental effects on the environment.

Water Scarcity

The apparent abundance of water is deceptive and we tend to take it for granted. We tend to abuse and overuse it. This has led to water scarcity, for which the reasons are:

- Increasing demand for water due to rapid increase in population.
- Bad management: India receives the second highest amount of rainfall in the world, next only to Brazil, almost 1150 mm. annually, draining a total volume of 400 million hectare metres. Of this, only 100 million hectare metres is retained in the soil. The rest is carried as run off into the oceans and seas.
- Poor ground water resources due to deforestation and overgrazing which result in soil erosion and inability of the soil to permit water infiltration.
- Poor storage facilities.
- Over-exploitation of groundwater in water-scarce areas often resulting in the intrusion of saline sea water in coastal areas.
- Dumping of various types of pollutants into our water bodies, reducing the usability of the available water.
- Cultivation of hybrid varieties of paddy, wheat, cotton, sugarcane and tobacco which consume more water.
- Evaporational loss of water stored in large reservoirs and loss by seepage in long canal systems.
- Siltation of bodies of water due to denudation of the vegetational cover in the hills and catchment areas.

In many parts of our country, water is scarce. In villages, women have to walk several kilometres everyday to collect a few pots of water, sometimes making several trips a day. Even in cities and towns, people face great hardships to collect a few buckets of water. With our ever-multiplying population, water scarcity is going to increase, not diminish. It is time we used our resources wisely.

Water Conservation

- Water should not be wasted. Leaky taps must be repaired and taps closed when not in use.
- Rain water can be collected and stored for domestic purposes.
- Delivery lines must be maintained properly and faulty and leaking pipes replaced. Plumbing should be done correctly, seeking advice from experts.
- Hand pumps should be well maintained to perform effectively.

- High HP motors to draw water from the well (open or bore) must not be used when low HP pumps can perform the same function.
- Overhead storage tanks should be maintained in good condition. PVC tanks which are cheaper and lighter in weight and which will not corrode or rust are preferable to metal tanks.

Management of Water Resources

Desiltation

- Canals, tanks, yeris, etc. must be desilted regularly during the summer months.
- People should be encouraged to revive the ancient practice of protecting trees around tanks. Major rivers were sanctified and groves were established around villages and on the river banks.

Afforestation

Afforestation of barren, hilly slopes on a warfooting should be carried out. Trees withstand drought better than crops. They check dust, replenish streams, provide shade to cattle and man and give fodder for cattle. They provide innumerable uses for man. Denuding the land of trees without compensatory afforestation is a suicidal and short-sighted approach to solving immediate needs.

Ponds and Tanks

- Creation of small reservoirs and percolation tanks to hold run-off water must be implemented and maintained well.
- Agronomic practices like off-season tillage (prior to pre-monsoon showers) conserve soil moisture. Moisture penetration to a depth of 90 cm. is achieved if the land is ploughed to a depth of 30 cm. Other practices like early sowing of seeds, moderate use of fertilisers, weeding, pest and disease control and timely harvesting increase the yield inspite of limited moisture in the soil.
- Terrace cultivation of hilly slopes prevents water run-off.

- Contour ploughing and planting of grasses and trees check run-off water and increase the soil's capacity to retain moisture.
- Green manuring (incorporation of fresh green leaves into the soil) and crop-rotation (cultivating different crops in rotation depending upon the soil and climate, e.g. cereals followed by legumes) conserve soil moisture.
- Mulching the soil with organic residue conserves soil moisture.
- The use of sprinkler irrigation for closely-spaced crops like millet, pulses, groundnuts, etc., conserves 30 to 40% of the surface water.
- Drip irrigation is most suited for closely-spaced row crops like vegetables, cotton, sugarcane. The efficiency of this system is around 25 to 30% in conserving soil moisture. The cheapest and easiest form of drip irrigation is to drill one to three holes in a mud pot and bury it partially in the soil next to the plant. The water in the pot drips slowly, ensuring that the soil is continuously moist and the plant gets a constant supply of water.
- Harvesting rain water and storing it in small ponds ensures water supply during summer.
- Deep trenches can be dug adjacent to bunds to collect run off water and soil.

Water Pollution

Like clean air, fresh water is also becoming a scarcity. The limited availability of fresh water and its unequal distribution make water pollution a matter of great concern. Water pollution is generally localised and confined, making it more severe. The pollutants undergo many reactions and can become hazardous. 70% of India's fresh water is polluted, including several high altitude lakes. While water pollution is easier to study and manage, its control is highly complex and very costly.

Sources of Water Pollution

In underdeveloped countries, sewage is a major source of water pollution. Human excreta contains 400 different species of bacteria and viruses. Even well-treated sewage contains pathogenic bacteria and virus, unless properly chlorinated before being discharged into any water course. Sewage is a major contributor to water-borne diseases and affects the health of people and other organisms in the environment in many ways.

Industrial effluents from sugar factories, distilleries, tanneries and paper industries are accompanied by very high organic loads. By-products of paper and pulp industry cause depletion of fish upto as far as 40 km downstream. The wastes from oil refineries and steel industries contain phenol which imparts a strong odour, apart from poisoning the water body. Fertiliser industry wastes contain ammonia, urea, phosphate and sulphate which, in water, cause algal bloom and are toxic to aquatic fauna and flora. Alkaline industry wastes contain mercury which can kill human beings who consume mercurised fishes. Lead generated from battery, printing, petrol and paste-processing industries, trace and toxic elements such as zinc, copper etc., and effluents from mining industries are injurious to aquatic organisms.

Water in which maximum permissible concentration of any single or more constituents is in excess is unfit for drinking and human health. There are definite tolerance levels for water used for different activities such as drinking, bathing, irrigation and industrial purposes. Depending on its use, there are different physio-che .iical and bacteriological standards for water.

Major water pollutants and their effect on human health.

There are some ways in which bodies of water get polluted:

- Discharge of untreated domestic sewage into waterways.
- Discharge of industrial effluents.
- Discharge of agricultural wastes into waterways.
- Use of pesticides, herbicides, fungicides and weedicides which seep into the ground water.
- Discharge of fertilizer residues as run-off from farm lands into waterways.
- Accidental spillage of oil in the seas by tankers.

Control of Water Pollution

- Legislation should be enforced to control water pollution.

- Effluents from industries should be treated and then discharged into water bodies.
- Treated effluents can be recycled for other uses like agriculture, and the sludge after drying can be used as manure.
- Many more sewage treatment plants must be installed. At present, raw sewage is fed directly into water bodies.

Water Purification - At Home

- The vessels which store water should be clean.
- Rusted, greasy vessels should not be used for storing water.
- Water must be filtered through a clean, white cloth before storing.
- The traditional 3-pot filters can be used for purifying water.
- Water must always be boiled before drinking.
- Modern water filtering equipment capable of removing bacteria can also be used.

Water Purification - by the Community

- The surroundings of water bodies should be kept clean.
- Care should be taken to prevent the disposal of sewage and domestic or industrial waste into bodies of water.
- Proliferation of Eichornea (water hyacinth), Pistia and algal bloom on water must be avoided.
- Fish should be introduced into bodies of water to purify the water.
- Wells should be kept open to allow fresh air and sunlight to act as a natural sterilizing agent on the surface of the water.
- Wells and tanks must be desilted regularly.
- Washing of clothes and vessels, cattle and automobiles should be prohibited at public tanks, yeris, etc.

Water Purification - by Industries

Effluents must be recycled and effluent treatment plants must be installed. Each large industry should have its own water treatment plant and small industries should contribute to a co-operative plant.

Water pollution by industrial effluents must be treated at the site before the effluents are released into the water body. The Government should punish industries heavily if they pollute fresh water, and should supervise and inspect industrial effluent treatment plants to ensure proper de-toxification and recycling measures. LEGISLATION Right to life, which is a fundamental right under Article 21 of the Indian Constitution, has been interpreted by the courts to also include a right to pure air and water.

Time and again many citizens avail of this provision to fight against polluted water.

Section 277 of the Indian Penal Code deals with water pollution, which reads as follows:

Whoever voluntarily corrupts or fouls the water of any public spring or reservoir, so as to render it less for the purpose for which it is ordinarily used, shall be punished with imprisonment which may extend to three months, or with the fine of one thousand rupees, or with both.

The more recent legislation on water pollution is the Water (Prevention and Control of Pollution) Act, 1974. This Act is meant to curb the various kinds of pollution ranging from domestic to industrial pollution. Violations under this Act are more severe.

Parts X-B and XI-A of the Merchant Shipping Act, inserted by the Amending Act of 1983 deals with every aspect of marine pollution.

Note: Any citizen, who finds any water source polluted by a government or private agency may give a complaint to the State Pollution Control Board (PCB) and wait for 60 days for the PCB's action. Only in cases where the PCB does not respond can the citizen go to the court. However it is to be noted that this applies only to suits under the Water Act of 1974.

Another important right available to the citizens of this country is that where there is no water available for them to sustain their life and if that is caused due to the neglect of the State, that individual or individuals can file a writ petition either under Article 32 or Article 226 of the Constitution for the violation of their fundamental right, i.e. right to life which is enshrined under the Article 21 of the Constitution. This could also be applied to cases where the drinking water source is polluted by a government or private agency.

8

India's Environmental Policy, Programs and Politics

In June 1972, Mrs. Indira Gandhi emphasized at the first U.N.-sponsored conference on environment that poverty was the worst form of pollution, and consequently, it was a most urgent issue facing the international community. During the next two decades, her prophetic words were not heeded, while poverty continued to engulf the Third World. This issue was then again highlighted by the World Commission on Environment and Development Report, *Our Common Future*. At the Earth Summit in Rio de Janeiro, the removal of poverty as an integral part of the world strategy for environment and development became a truism, although still lacking commitment from most of the industrialized nations.

Between the period from the United Nations Human Environment Conference in Sweden in 1972 to the Earth Summit in Brazil 1992, India took various steps to control pollution and conserve resources. It also kept reminding the industrialized world that so long as poverty remains the main stumbling block in its road to development, its efforts to protect the environment and conserve resources would not bear the necessary fruits. India's sentiments have been echoed by other Third World nations. For India, as for other nations of the South, removal of poverty and environmental protection are the two sides of the same coin, sustainable development.

During the two environmental decades (1972-1992), the major issue before the government of India has been how to strike a balance between the urgent need to secure food sufficiency, health, drinking water, and shelter and the desire to protect the environment. Although various institutional and legal mechanisms were initiated, the pressures of the growing population and an ineffective system of environmental law enforcement, along with the scarcity of funding, have added to the misery of the nation. The government of India was forced to acknowledge through its "Policy Statement for Abatement of Pollution" that "the state of the environment continues to deteriorate." India's response to environmental problems is being watched by nations the world over – because if India is able to prevent the deterioration of the environment and provide for the removal of poverty, it will be an example for others in the Third World to follow. In this chapter, the authors have attempted to examine the challenges and responses to the environmental problems faced by India.

Emergence of Environmental Concern and Development of National Policy

The major environmental concerns of India today are the result of continuing poverty, coupled with growing population and the side effects of enhanced industrial activities (including human settlements and movements). Conditions of poverty affect the meager natural resources which the nation wishes to protect and conserve. They also create problems with respect to the fulfillment of basic human needs for a large section of India's population. At the same time, India's people are impatient when it comes to making attempts to improve their living conditions. Towards that they would not hesitate to put more demands on the environment if the short term gains could ensure basic daily needs. Population has become a major source of environmental degradation because it now exceeds the threshold limits of the life-support system within the country's boundary. The population of 844 million in 1991 is increasing at the rate of 2.11 percent annually, adding about 17 million people. In addition, India has probably the largest cattle population on earth, about 500 million head, but only 13 million hectares for grazing. The multiplying population of both the human beings and animals is putting a tremendous pressure on the environment. In the race for survival, both animals and human beings suffer. For example, over 250 million children, women, and men suffer from under nutrition. The prospects for the future are alarming indeed. In addition to the distressing situation arising out of population pressures brought by human beings and animals. India faces immense challenges from the side effects of industrial development and human settlements; some of these are highlighted here:

1. Dwindling forest cover caused by overgrazing, harvesting of trees for commercial and domestic fuel purposes, and illegal encroachments. During 1987-88, the actual forest cover in the country was 64.07 million hectares. Most of the loss has occurred since independence. This has not only resulted in the extinction of rare plant, animal, and microbial species but also contributed to soil erosion, floods, and groundwater depletion.
2. Polluting industries: There are about 4,000 water-polluting large industries, with only half of them having any pollution abatement plants installed. There are over one million medium-sized industrial plants that do not have any pollution-abating units. These industries have polluted the river basins and groundwater to such an extent that in the case of the holy River Ganga, the government of India had to establish a special authority to monitor the water pollution-causing industries and to require them to install abatement units, as well as to require the local municipal authorities to install sewage treatment plants.
3. Air pollution: In many cities, industries are located too close to the residential areas. Emissions from these industries are causing many respiratory diseases, aggravated by auto emissions and population congestion. The severity of the Bhopal disaster could have been less had it not been for the proximity of the Union Carbide Plant to the city population.
4. Unplanned and encroaching urbanization is yet another environmental pollutant. India is actually facing a classic urban nightmare. In 1981, there were only 12 cities with populations over one million. By 1991, the number had climbed to 23. In 1951, India's urban population was 62 million; by 1981, it had grown to 158 million; and by 1991, it

reached about 227 million; among these, over 30 percent of the population live in slums. When such a large number of people is concentrated in small, unplanned spaces, they suffer not only from the shortage of drinking water and sewer facilities but also from a host of other diseases.

5. Environmental refugees: India has undertaken to build a large number of mega projects related to dams and to mining and mineral exploration activities. These projects have caused a much less talked about environmental refugee problem. A study reveals that about 14.5 million people have been displaced, and among these only 3.9 million have been rehabilitated. Plights of these people have attracted worldwide attention; one example is the reversal of the decision in 1992 by the World Bank not to finance the Narmada River Project.
6. Pesticides and agriculture: India is the largest producer and consumer of pesticides in South Asia. The most productive agricultural regions of the nation show a severe pesticide pollution problem (including the use of insecticides, herbicides, and fungicides in the country). "More than 665 people died in 1989-90 because of pesticide poisoning, and thousands of cases of crippling resulting from pesticides have been recorded in the country." In addition to problems arising out of the indiscriminate use of pesticides, another worrisome problem is the fertilizer pollution resulting from the intensive farming methods used in various parts of the country. Synthetic inorganic nutrients generally escape to river basins, dams, and coastal waters causing algae growth, oxygen depletion, and related problems.

The year 1972 marks the watershed in the history of environmental policy development in India. Prior to 1972, environmental concerns, such as sewage disposal, sanitation, and public health, were dealt with by different ministries of the government; and each pursued its objectives without any integral coordination either within the federal structure or between the federal and state governments. Prior to the convening of the World Conference on the Human Environment in Sweden in 1968, the U.N. General Assembly requested each country to prepare a report on the state of its environment. India established a National Committee on

Environmental Planning and Coordination (NCEPC) in February 1972. The committee not only prepared the country report, which was presented by Prime Minister Indira Gandhi at the Swedish conference in June 1972, but also started establishing linkages with the Planning Commission and other federal ministries, as well as state governments. An Office of Environmental Planning and Coordination was established to assist the committee with the federal Department of Science and Technology. This office became the precursor of the federal Department of Environment and eventually the Ministry of Environment and Forests.

Between 1972 and 1980, NCEPC tried to establish its functional relationship with other power ministries within the government but faced problems mostly because "different departments started viewing this committee as a competitor rather than planner and coordinator." Despite these two problems, the efforts of NCEPC paid off. The Fifth Five-Year Plan incorporated environmental aspects into the planning and development process. A constitutional amendment in 1976 obligated the nation to protect the environment. Article 48A of the Constitution Amendment declared that the state shall endeavor to protect and improve the environment and to safeguard the forests and wildlife of the country.

This constitutional commitment was the major step taken by any nation on earth as India became the first country to change its constitution to protect its environment. However, between 1977 and 1979, not much happened in the field of environmental protection, partly because Mrs. Gandhi was defeated in the 1977 general election. For the Janata party, environmental protection was not a priority item; but the January 1980 general election brought Mrs. Gandhi back into power. She appointed a committee under the chairmanship of N. D. Tiwari to recommend legislative measures and appropriate machinery for environmental protection at the federal level. At the recommendation of this committee, a separate Department of Environment was established on November 1, 1980. It was hoped that state governments and Union Territories would also establish such mechanisms.

The Sixth Five-Year Plan (1980-85), devoted an entire chapter on environment and development, emphasizing sound environmental and ecological principles in land-use planning, agriculture, forestry management, wildlife preservation, water and air pollution control, marine and coastal zone protection, fisheries, renewable resources and energy, and human settlements. The plan also provided guidance on environmental concerns to the federal and state administrators; and further, it allocated financial resources for environmental conservation and protection efforts.

Various changes also took place in the institutional setup at the federal level. Being the focal point for environmental policy planning and management it assumed responsibility for environmental planning, coordination (not only among various federal government ministries but also between the center and state governments), environmental awareness and promotion, preparation of environmental policy guidelines for various sectors, environmental appraisal of development projects, promotion of environmental research and education, and international cooperation. In order to accomplish these activities, the department brought together a multidisciplinary team of scientific, technical, and administrative professionals in its secretariat. Several scientific agencies, such as the Zoological Survey of India, the Botanical Survey of India, and the National Museum of Natural History, were brought under the authority of the department. In January 1985, the government created a Ministry of Environment and Forests so that the various aspects related to natural resources (under the legislative control of the federal government) could be brought under one administrative setup and an appropriate coordination could be achieved. Thus, upgrading the status of the department provided added clout for the agency to press for pollution control and environmental conservation activities. At the end of the 1980s, the status of this ministry was upgraded. In addition to a minister of state, a senior minister with the rank of cabinet minister was appointed. The passage of the Environmental Protection Act of 1986 enhanced the policy and institutional base of the ministry.

Since its inception, the ministry has undertaken a massive pollution-control program. In February 1985, it established the Central Ganga Authority (CGA) to improve Ganga River quality by reducing the pollution load, establishing a series of sewage treatment plants along the main urban centers, renovating all the existing sewage pumping and treatment plants, providing waste-water sub-pumping stations at the outfall points of open drains not yet connected to the existing sewage system, extending the sewage treatment system to cover the unsewered areas, and constructing an electric crematorium. By 1992, 261 schemes were sanctioned by the Central Ganga Authority; and of these, 173 had been completed. Seven new sewage treatment plants, out of a total of thirty-five to be established, have become operational. The authority has also identified 68 industries along the river basin that are grossly polluting the river. Using the provisions of the Environment Protection Act of 1986 and the Water Pollution Prevention Act of 1974, the authority has been able to get 43 of these industries to install their effluent treatment plants. Eight of the remaining 15 are constructing these units. Ten industries have been legally closed, while negotiations are under way with 7 of the industries which do not have the effluent treatment plants. Other major initiatives undertaken by the ministry during the decade 1980-1990 are survey of natural resources (including intensive floral survey, faunal studies and survey, forest survey), passage of the Wildlife Protection Amendment Act 1991, establishment of tiger reserves and biosphere reserves, creation of guidelines for environmental impact assessment for major thermal and hydroelectric power projects, formulation of a national river action plan to undertake works (akin to the Ganga action plan) in other grossly polluted stretches of major rivers of the nation, establishment of the National Wastelands Development Board to promote afforestation and wastelands development, funding for various research studies (dealing with wetlands, mangroves, biosphere reserves, forestry, wildlife, and items relating to the Ganga action plan), environmental education and forestry-officers' training programs, wildlife education and training program, provision of cooperation to international programs, and participation in the formulation of international conventions and treaties, such as the Global Convention on Conservation of Biodiversity and the Montreal Protocol. In order to accomplish these activities, the size and budget of the ministry have grown. In 1982, the Department of Environment had 250 (including technical and administrative) personnel with five divisions. By 1992, the number had grown to 1,023, with 23 divisions and various directorates and boards, such as the Ganga Project Directorate and National Wastelands Development Board, thirteen autonomous agencies (such as the Central Pollution Board in New Delhi, Center for Mining Environment), sixteen associated units (such as Forest Survey of India, Zoological Survey of India, Botanical Survey of India, Wildlife regional offices, and National Museum of Natural History), and one public-sector undertaking, Andaman & Nicobar Island Forests & Planning Development Corporation Ltd. During this period, the budget has increased from Rupees 138,643,000 to Rupees 2,664,400,000.

Policy Instruments Used: The government of India, through its water pollution prevention and air pollution legislation has favored the policy of regulation and enforcement rather than the effluent charge strategy. The first mechanism requires the following four steps to be undertaken by the government:

1. The necessary national standards for the abatement of pollution must be established, which would then be implemented and enforced by the central and state pollution control boards. The regulatory mechanism also includes the introduction of an "environmental audit" program for industries, as well as government public enterprises and local government undertakings, so that the operations and activities affecting the environment, particularly with the compliance with standards relating to effluent and emission control, can be measured and evaluated. The Policy Statement for Abatement of Pollution issued by the government of India in February 1992 mentions the need for such an audit statement to be issued by industrial concerns and public undertakings. The government plans to change the existing standards, which are based on the level of concentration of pollutants in effluents and in emissions; instead, standards will be source related to encourage the minimization of waste and promote recycling and reuse of materials.
2. The government needs to establish a set of penalties for noncompliance with the standards and regulations. For example, the Water (Prevention and Control of Pollution) Act of 1974 provides for imprisonment for up to three months or a fine of up to Rupees 5,000 or both; in the case of continuing violations, an additional fine of Rupees 1,000 per day can be levied. Under the Air (Prevention and Control of Pollution) Act of 1981, imprisonment up to three months and a fine of up

to Rupees 10,000 or both can be handed down; in case of repetition, a fine of Rupees 100 per day can be levied; under the Environmental Protection Act of 1986, imprisonment of up to five months, a fine as high as Rupees 100,000, or both could be given; and in case of continuing offenses, a fine of Rupees 5,000 per day can be levied.

3. The third step involves a continuous process of monitoring and surveillance of the targeted industries to detect noncompliance with standards.
4. The fourth step requires that the enforcing agency take immediate steps, after warnings and recommendations have remained unheeded, using the judicial process to impose sanctions and penalties.

Legal Mechanisms

More than two dozen laws protect India's environment. They cover all aspects of the environment—from pollution to conservation, from deforestation to nuclear waste. Some of these laws existed even before the environmental movement arrived on the world scene: Forestry Act of 1927, Orissa River Pollution Act of 1953, Punjab State Tubewell Act of 1954, West Bengal Control of Water Pollution Act of 1957, Jammu and Kashmir State Canal and Drainage Act of 1963, and the Maharashtra Water Pollution Prevention Act of 1969 are just a few examples. Despite these laws, by 1970 India was faced with an environmental crisis. Immediate and drastic actions were needed to cope with the magnitude of this crisis.

In the following section, five major environmental laws passed since 1970 are identified. These laws are significant for two reasons. First, they help us understand the evolutionary process. Each major enactment points to a different set of political calculations and a different interpretation of costs and benefits. Second, the five laws form the nexus of all environmental activities in the nation. Despite being specific in nature, these laws together have built a common context for all environmental policy actions. Vogel calls this context "national style" and claims that different styles are the result of different underlying political processes and can lead to varying degrees of performance.

The Water Act of 1974

The Water Act was passed in 1974 for the prevention and control of water pollution. The inclusion of "water pollution" as an agenda item came in 1962. The first response came from the Ministry of Health which appointed an expert committee to study the problem and prepare draft legislation to regulate the quality of water. It was realized that the issue could no longer be handled by state governments, which, according to the Indian Constitution, are responsible for regulating water quality. The committee recommended a shift in jurisdiction from state governments to the federal government. In 1965, a proposal was sent to all state governments to relinquish their jurisdiction in this area and to authorize the federal government to legislate. The response was apathetic. After much persuasion, six states granted their consent in 1969, and draft legislation was introduced in Rajya Sabha, the upper house of the parliament, in the same year. However, the parliament failed to show much enthusiasm. The future of the issue remained ambivalent until after Indira Gandhi's return from the 1972 United Nations Conference on Human Environment. Mrs. Gandhi criticized the idea of channeling resources for the protection of the environment when nearly half of the Third World's population was hungry. However, upon her return, she assumed a crusading role in the promotion of the environmental issue, and the Water Act was finally passed overwhelmingly in 1974.

The Water Act covered all major sources of pollution by defining water pollution as "such contamination of water or such discharge of any sewage or trade effluent or of any other liquid, gaseous or solid substance into water as may, or is likely to create a nuisance or render such water harmful or injurious to public health or safety, or to domestic, commercial, industrial, agricultural or other legitimate uses, or to the life and health of the animals or plants or of aquatic organisms."

Under the legislation, a separate bureaucratic network was created-one central board, one state board in each consenting state, and some joint boards of two or more states to ensure abatement of interstate water pollution. The central water board was granted the right of advisement, execution, consultation, and coordination. State boards were granted the same rights within their territorial jurisdictions. Boards' powers were backed by several judicial provisions. Failure to comply was punishable by a term of three months in prison and a fine of Rupees 5,000. Repeated offenses could lead to a fine of Rupees 1,000 a day and a prison term of up to six years.

In later years, the provisions of the Water Act were strengthened twice. In 1977 the Water Cess Act was passed, and in 1988 the Water Act itself was amended. Under the Water Cess Act, each industry and local body was required to pay a nominal cess (tax) on water consumption. This cess was refundable up to 70 percent if the discharge of water did not exceed the permissible limits. Under the 1988 amendments to the Water Act, all industries—new or old—were required to obtain environmental clearance. In addition, penalties were doubled, and powers of the boards were more specifically defined.

While the Water Act provided comprehensive coverage of the issue of water pollution, it left many grey areas difficult to administer. First, the scope of the act was inadequately and restrictively defined. It failed to cover some critical sources of water pollution, such as groundwater contamination. Second, it allowed government agencies an undue flexibility. Municipalities, which should be primarily responsible for treating residential waste, remained free from direct liability. The head of any agency was not punishable "if he proves that the offense was committed without his knowledge or that he exercised all due diligence to prevent it."

Third, the act failed to establish an independent revenue base for the boards. The Water Cess Act was an attempt in this direction; nonetheless, the amount of the fee was too small to make any dent. Fourth, the act did not allow victims the right of prosecution. Charges could be brought to the courts only by boards. However, boards were overburdened and under funded to such an extent that only 5 percent of public complaints could be processed by them. Fifth, the penalties established by the act were also unrealistic. The cost of defiance for polluting industries was lower than the cost of compliance. Consequently, industries found it cheaper to pay fines than to adhere to pollution control guidelines. Last, the act provided no mechanism to prevent board "capture." In fact, during the last decade water boards have faced pressures from two sources: first by regulated interests; second by political bosses.

On the whole, the performance of the Water Act remains dismal. The level of pollution has not declined, although it is difficult to guess what would have been the level of pollution, had there been no legislation. According to the National Engineering and Research Institute (NEERI), 70 percent of the available water is polluted, and an estimated 73 million workdays are being lost every year, costing society 60 million Rupees 6 crores, because of pollution-related health problems. In addition, only 3 percent of the total population has access to the sewage treatment system and only 27 percent of the large and medium-sized industries have full or partial treatment facilities.

The Air Act of 1981

The problem of air pollution is as severe as the problem of water pollution. According to a survey, the level of air pollution in major cities has nearly tripled in the last fifteen years. In 1984, 31 out of 48 thermal plants had no pollution-control measures. Over 40 percent of the vehicles in Delhi did not comply with emission standards. According to the 1991 World Resource Institute, India ranks as the fifth largest climate polluter in the world, ranking after only the United States, the Soviet Union, Brazil, and China. Motor vehicles provide half the pollution in major cities. The countryside is not spared pollution either. Rural air pollution is caused primarily by wood-burning stoves. It is estimated that in the three hours of cooking time, women inhale as much carcinogenic benzopyrene as is contained in a pack of twenty cigarettes.

To control air pollution, the Air (Prevention and Control of Pollution) Act was enacted in 1981. The legislation defined air pollution as the presence of any "solid, liquid, or gaseous substance in such concentration as may be or tends to be injurious to human beings or other living creatures or plants or property or environment." Like the Water Act, the Air Act specified penalties and prosecution guidelines. Also, no new industrial outlets could be built without the prior consent of air boards.

The Air Act suffered from the same shortcomings as the Water Act. Its language was not all-inclusive, penalties were minimal, time limits were not set, public input was not sought, and third-party rights were nonexistent. In addition, administrative support for the act was also questionable. Although the legislation provided for the creation of a separate enforcing network, the enforcement was placed under the jurisdiction of the water boards. However, water boards were already resource starved and, in reality, could not cope with the burden of new responsibilities.

The Environment Protection Act of 1986

A small gas leak from Union Carbide's plant in Bhopal caused a major environmental disaster in 1984. The incident was labeled an accident, and blame was placed squarely on Union Carbide. Nonetheless, the incident shook both the world and the Indian policymakers. It was realized that despite fifty pieces of legislation, thirty of them on pollution alone, society was not safe from environmental disasters. This realization led to the enactment of the Environmental Protection Act (EPA) in 1986. The act has three major objectives: (1) to be supplementary rather than a new law, (2) to provide a clear focus of authority, and (3) to "plug loopholes" in existing laws. The act provided the first official definition of "environment" and emphasized the interrelationship between people and their environment: "'Environment' includes water, air and land and the interrelationship which exists among and between water, air and land, and human beings, other living creatures, plants, micro-organism and property."

The Environmental Protection Act allowed the central government to delegate authority and to use existing resources to implement the act. The act was similar to the Water Act in tone, words, and emphasis. Nonetheless, there were some added features. The act covered the entire nation and was not limited to the consenting states – a feature of the Water Act. Penalties were doubled. In addition, the third-party right to prosecute was granted, even though to a very limited extent. The act stated that any person, after giving a notice of sixty days to the central government, may take an offending party to the court. The act fell short of environmentalists' expectations because it barred civil courts from entertaining environmental complaints, it freed agency heads from assuming full responsibility, and it denied citizens the right to sue the government for ignoring its duties. According to Ranbir Singh, "This act has been likened to a cobra that is seemingly fierce. It raises its hood and hisses menacingly but if you pry its jaws open, you will discover it has no venom in its fangs."

The Forest (Conservation) Act of 1980 and National Forest Policy of 1988

Forests in India received special consideration from the British. They were used for export purposes, and their protection was considered crucial to the British interests. In 1878, a special ordinance passed by the British classified Indian forests into three categories: restricted forests, which were restricted from the people; protected forests, which were accessible to the people although certain restrictions applied; and village forests, which were open to the public. This classification, for the first time, defined forests as commercial goods and placed them directly under the government's domain.

In 1927, the British passed the Indian Forestry Act and took complete control of the forests. Since then, the management of forests has been directly under the control of the government. The government of free India accepted this responsibility and passed the Forest Policy Act in 1952. At the time of the policy's inception, 22 percent of the nation was under tree cover. One of the goals of the policy was to increase this cover to 33 percent. However, several reports in the late 1970s indicated that instead of increasing, the tree cover had, in fact, been further reduced. Consequently, the government of India issued a guideline to all state governments against the use of forest land for any nonforestry purpose. Unfortunately, the guideline proved ineffective. Within the next 25 years, tree cover was reduced to only 14 percent of total area. Consequently, the government passed the Forest (Conservation) Act in 1980. The passage of the act was primarily based on the truth that the forests are being destroyed very fast in the country, especially in the Himalayas and other hill areas. The continued process of ruthless destruction of forests is not only leading to heavy erosion of soil, erratic rainfall, and frequent floods but also causing acute shortage of fodder, industrial wood, and firewood. More significant is the loss of productivity due to degraded and eroded lands.

According to the Forest (Conservation) Act, a prior approval of the central government was required for any diversion of forest land by anyone, including government agencies. The definition of "non-forest purposes" included developmental projects, transportation projects, and cash crops, such as tea, coffee, spices, rubber, medicinal plants, and horticultural plants. The unique feature of the act was that it placed liability on public officials and in cases of negligence allowed their imprisonment for up to fifteen days. In 1989, nearly 4,000 requests were submitted, of which nearly 50 percent were approved.

The performance of the forestry laws left much to be desired. According to a report by the Forest Survey of India released in 1987, 1.5 million hectares of forests were still being lost each year; and 12,000 million tons of soil were being lost to erosion each year. Among other consequences, this assault on forests doubled the vulnerability of the surrounding land to floods and droughts in just one decade, from 1970 to 1980. Needless to say, environmentalists were beginning to question not just the design of the policy but also its underlying assumptions. Consequently, in 1988, revised version of the 1952 National Forest Policy came into existence. The revised policy declared seven basic objectives of forest management: (1) maintenance of environmental stability and conservation of the country's natural heritage; (2) checking soil erosion and sand dunes; (3) increasing the forest cover through social forestry; (4) meeting the rural people's needs for fuel wood, fodder, and timber; (5) increasing the productivity of forests to meet the nation's increasing needs; (6) efficient use of forest produce; and (7) creating mass awareness.

The newly stated objectives clearly indicated a more regulatory direction in the management of the forests. Increasing forest productivity, motivating people to develop forest areas, providing fodder and fuel through "conveniently located depots" at reasonable prices, and approving land use proposals all were to be the responsibility of the government. Even though the policy was applauded for recognizing the dependency of the people—rural and urban—on forests, it was also vehemently criticized for ignoring the need of decentralization and deregulation of the process. It was reported in a newspaper that "nothing has been proposed about the forest department, land laws, judiciary, panchayats and so on. If supplementary changes in socioeconomic and administrative legal system are not required, why a new policy at all?"

During the last two decades, the government of India has made significant progress in providing legal protection to the environment, yet several loopholes exist. For instance, noise pollution is not covered by any law, though several laws, including the Air Act, the Factories Act, and the Motor Vehicle Act, put certain restrictions on polluters. The use of pesticides is another overlooked area. According to a report, out of 22,000 deaths worldwide due to pesticide poisoning, about one-third take place in India. BHC, DDT, and endosulfane are the primary causes of pesticide pollution.

In addition, nuclear waste continues to pose significant problems. This waste is controlled under the Atomic Energy Act of 1962. Under the act, the federal government has full jurisdiction over the management of nuclear waste. The activities of the Department of Atomic Energy, which is the primary agency responsible for nuclear waste management, are not open to public scrutiny for national security reasons. This non-accountability keeps this policy area away from the active environmental agenda.

Institutional Support

Institutional support for the environmental issue began with the creation of the NCEPC in 1972. The move was a direct response to the call by the United Nations to prepare a status report on the nation's environment. NCEPC was an advisory body, and its scope was extremely limited. Nonetheless, it served an important function by initiating a forum for political debate on the issue. Water boards created by the Water Act of 1974 were the second major providers of institutional support. However, they too were ineffective. The trivial nature of the institutional support base changed in 1980 with the creation of the Department of Environment (DOE). The body was created in response to the recommendations made by an investigative committee (known as the Tiwari Committee) appointed by Indira Gandhi (Department of Science and Technology, 1980). Like its counterpart of the same name in the United Kingdom, the DOE was formed by amalgamating components from different departments and independent agencies. In its watchdog capacity, the department was to carry out several environmental functions, such as appraisal of developmental projects, protection of wildlife, monitoring of water and air quality, promotion of research, and international cooperation. The DOE faced harsh criticism right from its cradle. It had no political or financial base; therefore it failed to perform as a regulatory agency. It was ineffective partly because of its advisory capacity and partly because of its fragmented structure.

The Ministry of Environment and Forests

With the rise of Rajiv Gandhi as "the new age leader" in 1985, the issue of environment received some serious attention. Environmental laws were tightened, new programs were initiated, and the enforcing network was significantly strengthened. The major step taken by the Gandhi government was to raise the status of the Department of Environment to a ministerial agency. The Ministry of Environment and Forests (MEF) came into existence in 1985. Several organizations were brought under its fold, and several new components were created to give it a unified status.

The ministry today is headed by a union minister of environment, who is assisted by a minister of state for environment and a full-ranked secretary. It consists of eighteen divisions, eleven autonomous agencies, and six associated units. Since its inception, the MEF has continued to grow in size and specialization. Forest conservation was added as a division in 1987, and hazardous substances management and forestry international cooperation in 1988. In 1991, the ministry had a total technical and clerical staff of 1,023 in its headquarters in New Delhi. This number indeed is minuscule when compared to the staffing strength of 15,000 in the U.S. Environmental Protection Agency; nonetheless, it is a significant increase from the figure of 250 in 1982.

The ministry performs two types of functions: custodial and regulatory. Under the custodial functions, it assumes full responsibility over certain aspects of environmental management. These include pollution monitoring, ecogeneration, assessment of flora and fauna, forest resource development, wildlife protection, and wetland management. Under the regulatory functions, the ministry sets standards, conducts environmental impact assessments, and controls industrialization. One of its major regulatory programs is the Ganga action plan. Initiated in 1985, the plan is being implemented by the CGA, and aims to restore the quality of the Ganges. With a budget of U.S. $180 million, the plan consists of 262 schemes in the three states housing the river. As of 1990, 179 schemes were either complete or near completion.

The MEF is also entrusted with the right to conduct Environmental Impact Assessments (EIA). The objective of EIA is "to achieve sustained development with minimal environmental degradation and prevention of long term adverse effects on environment by incorporating suitable preventive and control measures." However, EIA, like many other regulatory tools, is proving ineffective. To begin with, the ministry is not properly equipped to carry out this major function. Lack of information and scarcity of basic resources have limited its ability to such an extent that the function itself has turned into a political burden. The MEF is criticized for delaying approvals beyond a tolerable time. Today, the ministry draws criticism not only from developmentalists and industrialists but also from its own constituents, which include social activists, pubic and elected officials, and state governments.

Ironically, the environment assessment function has created major conflicts even within the ministry itself. One side is claimed by technical experts who favor a purely technical assessment of projects, while the other side is supported by bureaucrats who prefer a more balanced look at them. In addition, the practice of EIA has also added to corruption because there is ample room for the use of administrative discretion at every level.

In addition, the ministry also serves as the primary source of data collection. The Environmental Information System (ENVIS) Network is housed in the ministry, with ten regional branches. The objective of ENVIS is to collect, store, retrieve, and disseminate environmental information to all concerned.

Institutional Support in a Federal System

One of the challenges of the environmental crisis is its complexity, which asks for global planning yet demands local action. Since all layers of government are involved in one way or another, environmental programs often turn into political turf battles. The Indian federal system is no exception. State water boards were the first exclusively environmental institutions to emerge. For a long time, they suffered from political malnutrition and economic disability. Kerala Water Board, for instance, claimed five years after its inception that "the board is still in its infancy and is working under severe handicaps of non-availability of funds to the extent required and consequent lack of manpower and infrastructure." It further presented its plight by saying that "the plan outlay under the retention and control of water pollution is fixed at Rs. 50 lakhs [approximately US $500,000] in the Sixth Five-Year Plan against the proposed outlay of Rs. 3 crores [$3 million] spread over a period of five years. But unfortunately the government has very recently suggested that even this meager plan outlay of Rs. 50 lakhs should be reduced to Rs. 30 lakhs."

In fact, the funding support of state-level environmental agencies remains a major source of discontent. In the budgeting process, a preferential treatment is given to the central government. In fact, the majority of the budget is at the disposal of the central government. Ironically, in the environmental field, while state governments are the primary implementers of environmental policies, they are not the primary receivers of environmental monies. The ratio of budget outlay for the Seventh Five-Year Plan (1985-1990) was 7 to 1 in favor of the central government. In addition to the water boards, 19 departments of environment also exist in the states. These departments are entrusted with the responsibility of providing assistance to their state governments and to implement policies of the Ministry of Environment and Forests. Constant appeals are made by the central government to strengthen the power base of these departments, but most still remain without adequate power bases. Their effectiveness varies from state to state depending on the states' commitments and their levels of public awareness.

In brief, the distribution of responsibilities—where the central government is responsible for making decisions and state governments for implementing them—has failed to work. The arrangement puts the state-level environmental agencies under the authority of two masters—the state governments and the MEF. Consequently, they face conflicting pressures—those set by the state governments and those set by the MEF. Often, the conflict emerges over difference in the calculations of costs and benefits. A state government may view a project as necessary for its economic sustainability and therefore may pressure the state environmental agency to be "reasonable" in its judgment. At the same time, however, the MEF may view the project from a national standpoint and pressure the state environmental agency to apply universal standards. State-level agencies are thus forced to please and represent two separate masters without having the resource bases to serve even one of them.

From Rajiv to Maneka and Beyond

While the issue of environment received attention during the Rajiv Gandhi years, this attention was secondary to Gandhi's overall emphasis on "technologization of India." After Gandhi's government failed to secure a majority in the 1989 elections, the nation came under the rule of a loosely assembled confederation of opposition parties led by V. P. Singh. From its inception, the Singh government was marred by disagreements among party members. The environmental cause remained untouched by this political warfare, however, primarily because of Maneka Gandhi. Singh appointed Maneka Gandhi, an administratively inexperienced yet environmentally committed leader, to head the Ministry of Environment and Forests. Maneka was aggressive and outspoken; her brash style of administration offered new opportunities to consider the environmental issue. She made it political simply by making it controversial. Her outspokenness annoyed public officials and businessmen alike and attracted media attention. Maneka Gandhi's strategy dealt with substantive issues. In brief, it combined the following items: (1) establishment of environmental courts to empower rural and urban people who thus far had no redress for environmental crimes committed against them, (2) comprehensive review of all environmental laws, (3) immediate crackdown on factory and vehicular pollution and implementation of the Motor Vehicle Act of 1989, (4) amendment of the Environmental Protection Act to make it mandatory for all projects to go through environmental impact assessment, (5) replacement of harmful chemical pesticides with biological pesticides, (6) introduction of the concept of environment-friendly products, (7) establishment of the National Waste Management Council, (8) revamping of Project Tiger, and (9) regular televising of environmental news.

One of the most controversial policy moves was the idea of establishing environmental courts to speed up the process of environmental decision making and make it more costly for polluters to pay the penalty than to comply. However, the proposal to establish environmental courts failed to win governmental support. By 1989, the Maneka agenda was beginning to become a political burden for the government. Her enthusiasm was unpalatable to a government that was growing politically unstable, as well as economically bankrupt. Consequently, a barrier was created in the person of Nilmani Routray, a 77-year-old political veteran. Routray was assigned to the newly created position of union minister, a position higher in authority than

Maneka's position as minister of state. Maneka was left with symbolic powers, which some newspapers referred to as nothing more than "the management of Delhi Zoo." Routray was given the ultimate responsibility to oversee the Ministry of Environment and Forests.

In 1991, Rajiv Gandhi's assassination changed the political equilibrium for the environmental issue again. Not only did the Congress(I) party lose its leader, it also had to survive on the support of other parties as it formed a minority government after the 1991 general election when neither Congress(I) nor any other party emerged as a clearcut winner. Consequently, a coalition government led by Congress(I) and headed by P. V. Narsimha Rao was formed. The Ministry of Environment and Forests was assigned to Kamal Nath. Kamal Nath led the India delegation to the Rio Earth Summit in June 1992. He reiterated Rajiv Gandhi's proposal that a Planet Protection Fund should be established by the rich North to assist the developing nations in their struggle for environmental protection and sustainable development. In the foreword to the national report prepared for the UNCED Conference in Rio de Janeiro, Kamal Nath stated: "It is clear that action for environment and development can be effective only if the various levels of government work in close harmony with NGO and concerned citizens." The ministry also issued a policy document in June 1992, the National Conservation Strategy and Policy Statement on Environment and Development.

While it is quite early to comment on the performance of the Nath ministry, some observations can be made. Overall, the support base for the environmental issue does not seem promising. Several factors account for this pessimism. First, a coalition government is hardly likely to take bold steps to shake up the existing equilibrium, particularly if public dissatisfaction is not threatening. Second, the troubled economy of the nation is unlikely to leave any room on the agenda for environmental issues. With the inflation rate in double digits and foreign debt nearing US $50 billion, the agenda is full of structural adjustments and economic recovery issues. Two issues are of particular importance—liberalization of the economy and streamlining of the public sector—both initiated in mid-1991 and both expected to have a lasting impact on the nation's priorities.

Environmental Policy and Management: Dangers Lurking?

Throughout the world, the feelings of desperation and failure are prevalent in the environmental policy area, partly because of the magnitude of the problem itself and partly because of an ever-increasing gap between policy goals and policy performance. Researchers offer several explanations for the environmental deficit. Most of these explanations can be clustered into two broad categories: policy related and administration related. Policy-related explanations point to the existence of a policy trap—or the absence of an adequate legal mandate to resolve the issue. On the other hand, administration-related explanations indicate the existence of an administrative trap—a situation where policies are ineffective because their administration is nonexistent or largely inadequate. According to Randall Baker, the administrative trap exists because "there is considerable dysfunction between the nature of ecological problems and the problem-solving structures within the public arena."

In India's case, the "policy-trap" argument can hardly be substantiated. Indeed, several areas of environmental protection fall in this category. Hazardous waste, nuclear waste, acid rain, and the ozone layer are clearly in need of more regulation. Yet there are sufficient environmental regulations to cover the environmental issue to a reasonable extent. The question is certainly not of numbers but of design. It is not the absence of a policy, but the inadequacy of its design that is at the core of the discussion because faulty policy designs lead to administrative traps which, in turn, lead to policy failure. The argument of administrative trap or administrative failure has already found wide support in the literature. These and related issues are discussed in the following sections.

Inadequate Foundational Support

While over two dozen environmental laws exist, they collectively fail to establish a solid, forceful, and comprehensive administrative network. More often than not, administrative guidelines are established without an effort to follow them up. The water and air acts are prime examples of this paradox. The administrative network created for the enforcement of the two pollution control laws fails on several accounts. It fails to establish any clear cut lines of accountability. Over centralization of legislative power cripples legislative innovativeness. Ironically, under the Indian federal system, only the responsibility for the environment is delegated, the authority to manage it is not.

Although the legal foundation for the environmental issue in India is a hybrid of the American and the British foundations, the mix is not properly balanced to suit the indigenous context of efficient management. In line with American policy, Indian environmental policy emphasizes adherence to standards, establishment of universal standards, and a centralized process, but it leaves the bureaucracy unaccountable. It does not create any mechanism to "watch the watchdog"—a function largely performed by the public and the media in the United States. Similarly, the Indian policy is an imitation of the British environmental policy when it comes to administrative flexibility and the use of discretion. Unlike the British policy, this flexibility does not properly reach to the local levels. This hybrid arrangement has produced an administrative authority that is inherently weak and enjoys little or no public support.

The Environment Protection Act of 1986 realized the need for an integrated policy approach, but even this act failed to create an integrated administrative network. Consequently, while the act filled policy gaps, it failed to fill the gaps in the administrative system.

Over the years, the environmental administrative machinery has successfully evolved from a showpiece agency into a viable force—from an advisory Department of Environment into a full-fledged Ministry of Environment and Forests. Nonetheless, it must still learn to use its force.

Lack of Administrative Rationality

The administrative network is embedded in a culture that does not foster compliance. Polluters and offenders find the cost of compliance greater than the cost of defiance. Low penalties and lengthy prosecution procedures make it easier for polluters to pay fines than for them to install pollution control measures. Only the sunrise industries, particularly the ones that enjoy the government's protection, can afford to install pollution control devices. For others, the competition to survive is so intense that it prevents them from making any new capital investments. In addition, even though the Ministry of Environment and Forests is developing a research base to provide technical assistance to the users of natural resources, the help is far less than the demand. At present, industries are burdened to find their own solutions. The cost of research adds another excuse for resistance. Several countries, notably Hong Kong, provide some form of institutionalized research support to polluters. The Hong Kong Productivity Council, for instance, helps industries by designing their individualized technological solutions. No such technological cooperation is available in India.

For regulators, the cost of enforcement is higher than the cost of nonenforcement. Public officials lack a political constituency. In India, they also lack public credibility. In addition to suffering from the lack of political support and public credibility, environmental regulators also suffer from interdepartmental rivalry. Relatively speaking, the MEF is a newcomer, entrusted with a role that infringes upon the traditional jurisdictions of other agencies, particularly of transportation, industrial development, rural development, and agriculture. Paradoxically, the MEF's power base does not match its complex set of responsibilities.

The politicization of the MEF further adds to the burden of regulating. Top positions in the MEF remain dependent on political loyalty rather than on environmental commitment—a phenomenon made clear by the appointment of Routray over Maneka Gandhi in 1990. Moreover, lower-level bureaucrats are not free from political and social pressures either. They are often undercut by powerful social interests. For example, it is a common practice for an industry to approach a higher-up "connection" if lowerlevel bureaucrats fail to adhere to their whims and fancies.

Lack of Grassroots Support

Even though significant social support for the environmental issue exists, opportunities to transform it into political support are minimal. Political parties prefer to rally around those issues that are politically attractive. Environment, despite its appeal, has a drawback: It has a counter growth dimension. As long as political parties support the principles of environmental protection, they remain in the safety zone. When they start to pinpoint strategies, such as industry closure or restrictions on forest contractors, however, they run the risk of being perceived as anti public. Thus, political parties, in general, shy away from micro level campaigning on the issue.

Interest groups also fail to provide political opportunities. Under the Indian parliamentary system, groups are forced to identify with political parties. Since voting in the parliament is on party lines, groups find it worthless to lobby individual legislators. Their interests are best served by keeping in touch with party bosses and with high-level civil servants. This situation forces groups to be captivated by political parties. Dependent on the mercy of the party bosses, the groups often turn to individual civil servants. Civil servants, thus, become the targets of "politicking."

India and the Earth Summit in Rio (1992)

The Indian delegation to the Earth Summit in Rio de Janeiro consisted of the Union Environment and Forest Minister Kamal Nath and nine ministers of environment from state governments. This delegation was preceded by a group of senior officials of the federal and provincial governments. The Indian delegation proposed a "Planet Protection Fund" to buy environmentally friendly technology and make it available gratis to any nation. This concept was first advanced by the Indian Prime Minister Rajiv Gandhi in 1989. The summit considered six treaties dealing with toxic and nuclear waste, foreign debt, forests, agriculture, plant and animal diversity, and a clearing house to share information on advances in environmental technology. In the summit, some nations were critical of the stand taken by the industrial nations. For example, the governor of the Brazilian State of Amazonia stated that the ecology movement was a First World plot to keep the Third World poor. The biodiversity treaty was signed by 98 countries, including Canada, India, and others, but U.S. President George Bush declined to sign it.

Concluding Observations

The two decades since the U.N. sponsored Human Environment Conference in 1972 have seen a steady and growing awareness among the elites in India about the ecological challenges facing their nation. This has been followed by an impressive growth of institutions established to deal with the problems of pollution and environmental conservation at both the national and the provincial level. At the same time, the mounting pressures of population, expanding urbanization, and the growing base of poverty have led to the ecologically unsustainable exploitation of natural resources which is threatening the fragile balance between ecology and humanity in India. Although India is not alone in this challenge, its problems will have to be tackled by its own people. The real challenge before India is how to preserve its environment, meet the basic needs of its growing population on an overburdened land, fulfill the necessary energy requirements of the people, and yet leave a legacy for future generations so that they may also enjoy the bounty of nature which the present generation is recklessly exploiting. The task is no doubt stupendous, for the hitherto unimaginative developmental strategies and largely imitative developmental schemes would have to be replaced by environmentally sound and sustainable development. The nation will have to search for ways to usher in *sarvodaya* in the least wasteful ways compared to the present system; for this, the country would do well to derive from its rich cultural heritage and the traditional conservation ethos a way to look ahead in harmony with environmental imperatives of the land.

9

Case Studies

NARMADA SAGAR DAM

Introduction

Social unrest at the grassroots level has shone both domestic and international light on the issue of dam projects within India. The Narmada Sagar dam (NSD) project and Sardar Sarovar dam (SSD) project in India have been the focus of widespread debate in recent years because of their potentially harmful implications. Dam projects undoubtedly take into account complex issues involving social, environmental, and economic factors. The decision to build a dam is a complexity in itself that should be analyzed to determine if the potential dam will have a positive impact in the overall view of development.

What are the social, environmental, and economic impacts of the NSD and SSD? Taking these impacts into account, are the projects beneficial to the development of the citizens in the dam's target area? A critical analysis of these questions may lead to potential modifications of these projects, or may even lead to a more critical stance when planning future dam projects. These questions will be answered through an analysis of the projects as they stand now, investigating their intended purposes and benefits. Social implications of the relocation of citizens in the areas to be inundated will be examined. The intended benefits of the projects for irrigation and hydroelectric power for citizens will be analyzed. Environmental consequences of the projects in terms of sedimentation, water logging, the destruction of forests, as well as water-borne diseases will be investigated. Finally, the economic implications of the projects will be examined, through the consideration of direct and indirect costs, and in addition to these factors, the involvement of the World Bank in these projects will be investigated.

The Narmada Sagar and Sardar Sarovar Dam Projects: A Description

A History of the Projects

The concept of building dams on the Narmada River originated from a British entrepreneur in 1863 who had the idea to harness the Narmada's water to irrigate crops in the drought prone state of Gujarat. The first technical studies for the damming of the river were conducted in the late 1940's. After more intensive studies, the Central Water and Power Commission came to the conclusion that the terminal dam on the river (the Sardar Sarovar dam) should be constructed at Navagam in Gujarat. It also recommended that the dam should be built in two stages, the first stage at a full reservoir height of 160 feet, and then the second stage upgrading to 300 feet.

In 1959 the proposed full reservoir height of the SSD was raised to 455 feet. This change occurred because it was found that water could be transported by canal to the Kutch region of Gujarat and eventually to the state of Rajasthan some 460 km away from the dam. The Narmada River flows through the states of Madhya Pradesh, Gujarat, and Maharashtra, and a dispute between these states arose over the distribution of the Narmada River's resources and the costs of the dam projects in the 1960's. With the purpose of resolving this dispute, the Central Government of India formed the Narmada Water Dispute Tribunal (NWDT) in 1969, and this tribunal sat for ten years. Thereafter in 1979, the NWDT awarded Madhya Pradesh 65% of the river's resources, 32% to Gujarat, 2% to Rajasthan, and 1% to Maharashtra (Dixon, Talbot and Le Moigne 1989, 54). Both the SSD and NSD projects initially did not meet the Ministry of Environment and Forests (MEF) regulations, and were both denied clearance for building in 1983. In 1987, due to political pressures of then Prime Minister Rajiv Gandhi's Congress party and the increasing damages from severe droughts in Gujarat, the SSD and NSD projects were given conditional clearance by the Central Government. Conditional clearance was given because, "the environmental assessment required under India's environmental regulations was not ready." The MEF allowed the projects to begin while the assessment was being completed, with the understanding that it would be completed by 1989, but as time elapsed, the assessment was not completed in 1989, and the construction continued.

Location and Description of the Dams

The SSD is situated near Badgam in the Bharuch district of Gujarat. The NSD is located upstream of the SSD, near Punasa in the East Nimar district of Madhya Pradesh. Both dams are the largest dams of the over 3000 dams proposed as part of the overall *Narmada River Valley dam project*. The NSD is designed to work in conjunction with the SSD by storing water in its reservoir for regulated release to the SSD. Water from the dam will be distributed to the states of Gujarat, Madhya Pradesh, Rajasthan, and Maharashtra as previously stated. In terms of physical structure, both dams are large-scale concrete dams being built across the Narmada river valley, with the SSD's full reservoir height set at 455 feet, and the NSD's height set at 860 feet.

What are the Intended Purposes/Benefits of the Dam Projects?

There are a few purposes for building dams in the Narmada river valley, and they include, "in the order of priority (but not in terms of volume of water use), the provision of drinking water, industrial water supply, irrigation and power generation". It is evident from the fact that in addition to the two dams, over 75,000 km of secondary canals will deliver water to irrigate farms, that irrigation is a major part of the dam projects. The command area of the projects (the area where water is to be delivered for irrigation in particular) is stated as being 2.0 million hectares in Gujarat (70% of which is drought prone), 3.1 million hectares in Madhya Pradesh, and less than 0.1 million hectares in both Rajasthan and

Maharashtra". Hydroelectric power is another major product of the dams. Dam officials have set the amount of hydroelectric power provided by the SSD at 1450 MW and 1000 MW from the NSD. Official also estimate that, "drinking water would become available for 3800 villages and 147 towns in Gujarat". The number of people who will benefit from the water and electricity delivered from the dams is estimated as being close to 40,000,000 with the majority being in Gujarat and Madhya Pradesh where a large portion of the population "is without adequate water or electrical supply".

Social Implications of the Projects

Who and How many People need to be Resettled from the Area to be Inundated?

It is estimated that over 250,000 people will be displaced by the SSD and NSD, and this have caused major dispute in recent years ("Damned" 1992, 36). Table 1.1 shows the breakdown of the number of villages and people affected by the two dams. The figures show that over 100,000 people will have to be relocated due to the SSD and about 170,000 people due to the NSD. The majority of the people affected currently reside in Madhya Pradesh, as both of the dams' reservoirs are located primarily in that state.

It is important to consider *who* is being relocated. The NSD's submergence area includes a population of over 67,000 tribal people. On the other hand, over two-thirds of those displaced by the SSD are tribal people. The fact that there is a large number of tribal people being displaced is significant because it has been observed that, "these tribals have been traditionally cultivating forest lands for decades [and] do not have legal title to the land: hence they will not get any compensation". Apart from the fact that the tribals may not get compensation, relocation will be detrimental to the livelihood of these peoples because they rely on the landscape for their beliefs, rituals, and festivals.

What are the Governmental Policies Regarding Resettlement in this Case?

Under the NWDT award given in 1979, oustees from any area submerged by the SSD have the option of being relocated to Gujarat or to stay in their own province. Unfortunately there is a lack of suitable land for oustees. Officials from Madhya Pradesh have offered financial compensation instead of land with the goal of sending oustees to Gujarat. Oustees in turn have refused to relocate in Gujarat stating that the land offered to them was "sandy and unirrigable", and thus unsuitable for them. As for those displaced by NSD, predominantly those from Madhya Pradesh, their only option is to relocate in villages within Madhya Pradesh. Again the problem of unsuitable land remains as, "the grazing lands that remain in the villages have hardly any soil cover and are of very poor quality, totally unfit for crop production".

In spite of the lack of suitable agricultural land for oustees, the government of Gujarat claims to have "the best resettlement and rehabilitation policy in the world for the people displaced by the projects". Under this policy, oustees are eligible for 2 hectares of land for each person, regardless of how much they had in the past, and temporary accommodations. This seems reasonable if it was actually put into practice with a respectable standard of living, but it seems that it has not. NGO's are claiming that oustees are not receiving their full share of land, and that the temporary housing consists of "tin sheds with one sheet of tin separating the family from their livestock". The government of Gujarat's resettlement policy does have its merits, though its implementation needs to be addressed in a more suitable manner. On the other hand, the government of Madhya Pradesh does not offer any relocation or compensation packages to "landless labourers who lose their livelihoods when their employers are resettled". If the government of Madhya Pradesh does not offer compensation to landless labourers then it is evident that *all* oustees are not being looked after.

How have the Citizens Affected by the Dam Reacted to Involuntary Relocation?

The World Bank's policy on involuntary resettlement says that, "displaced people should 'regain at least their previous standard of living'". Involuntary resettlement without suitable compensation also runs into the problem of backlash from the people affected. For instance, villagers in Jalsindhi, which is located in the area to be submerged by the SSD, are refusing to accept the rehabilitation package offered by the government of Gujarat. The villagers cite the problems of hostility of the local people in the areas of relocation, the lack of forests for wood, the absence of rivers for fishing, and the lack of pastures for grazing cattle. There are other instances where villagers are refusing to relocate and are fully willing to accept drowning as a form of protest. Public attention has increased as protestors and organizations such as the Narmada Bachao Andolan have been putting pressure on governments and capital lenders, such as the World Bank.

How have the Resettlement Activities to Date Affected the Construction of the Dams?

The Supreme Court of India ordered a halt on construction of the SSD on December 7, 1995. This order came about due to a lawsuit filed by the Narmada Bachao Andolan against the Central and state governments of India with respect to resettlement issues, environmental impact and economic non-viability. The State of Madhya Pradesh has proposed that the dam's proposed height be reduced to 436 feet from 455 feet before the case went to court, and now it wants the dam to be at a maximum height of 384 feet. The Madhya Pradesh government proposed the change in height because, "the largest number of oustees (80 percent) are from MP. It [the MP government] argued that the Gujarat government has not kept up its end of the deal because it was supposed to resettle people". Currently the SSD has been built to a height of 208 feet and the courts are refusing to allow construction to resume until authorities ensure the rehabilitation of the people displaced is secured. The reduction of the height of both the SSD and NSD is a critical issue because a reduction translates into less area being flooded. It has been estimated that if the proposed height of the SSD is reduced to 420 feet, 90% of the potential oustees would be eliminated, and 80% of the cultivable land to be submerged would be saved. In the case of the NSD, if the proposed height were reduced from 860 feet to 814 feet, the population displaced would be reduced to 20,000 people.

Are the Proposed/Stated Benefits of the Dam Projects Realistic?

When the NDWT awarded percentage shares of the water resources of the Narmada River and set the heights of the NSD and

SSD in 1979, it based its percentages and height on the quantity of water flow in the river. The tribunal used an estimate of 27.25 million acre feet (MAF) for the flow in 1979, using figures from 1891-1947. Using this figure, the tribunal calculated a full reservoir height of 436 feet for the SSD for irrigation purposes, and was later raised to 455 feet to allow for the generation of additional hydroelectric power for Madhya Pradesh. The initial figure of 27.25 MAF has now been deemed to be an overstatement, as records up to 1992 show an average flow of 23 MAF. With this figure, the height of the dam has been re-estimated so that it should be 400 feet high. Madhya Pradesh is willing to forego its share of hydroelectric power by reducing the height of the dam, because of the fact that most of the inundation is avoided at this height. India has had experience with overestimating figures in regards to dam projects. The Gandhi Sagar dam had overestimated figures and this left the top 5.4 metres of the dam dry, which translated into 30% of its capacity. A total of over 50,000 people were needlessly evacuated for the reservoir of this dam.

In the case of the NSD, the height of the dam was also set too high. This has a direct impact on the hydroelectric output of the SSD because the output was estimated assuming that the NSD would provide regulated releases of water for it at its proposed height. The government of Gujarat now says that the power generation of the dam will be closer to 425 MW instead of the stated 1450 MW during the first stage of the project before the irrigation system is in place. After the system is in place, the output will be reduced to a mere 50 MW.

The provision of drinking water is another benefit of the projects that has been stated by officials. Both dams do have the potential to provide drinking water, but "no plans have been completed for how the drinking water is to be delivered to consumer, nor has any money been allocated for this component of the project". Without plans or allocated funds for the drinking water component of the projects, it seems unlikely that drinking water will be provided by the projects any time in the near future.

Opponents to the projects are also pointing to the benefactors of the projects in their arguments against the dams. There is a belief that "the wealthy farmers, rather than the poor, 'will benefit most from the project'". The government of Gujarat has said that it will not allow cash crops to be grown in the command area of the projects, but again opponents point out that, "five of the largest sugar processing factories in Asia are now being built in the Sardar Sarovar project command area". There may be another reason why these factories are being built in this area, but there is a high likelihood that they are in anticipation of cash crops such as sugar cane.

Proponents of the projects point out that, "the region [command area of the dams] is 'one of the most severe drought-prone areas of the nation'," and that, "during a drought, bringing in water by truck or train and providing other basics costs about $200 million a year". It is true that the state of Gujarat experienced serious droughts in the 1980's and this was a contributing factor to the conditional clearance of the projects in 1987. The estimate that 40 million people will receive water and electricity from the dams, presented earlier in this paper, is one that should be considered when weighing the costs against the benefits of the project.

Environmental Implications of the Projects

Sedimentation and Siltation

Sedimentation is not a problem that dams cause, but is one that shortens a dam's useful life. It occurs when, "the larger sediments in the water entering a reservoir are deposited at its upper end forming a delta and steadily raising the level of the upper reaches of the reservoir." This sedimentation causes flooding due to its backwater effect, and in the case of the NSD and SSD it seems that, "no proper study of this has been done". Siltation also occurs as silt is deposited at the bottom of the reservoir, further reducing the dam's useful life. The estimated siltation rate for the NSD was initially set at 1.55 (hectares/metres of silt) per 100 km^2 of catchment area. From the table it is evident that all of the siltation rates have been underestimated. The same is true for the NSD, as observed surveys show a siltation rate of 5.62 (hectares/metres of silt) per 100 km^2 of catchment area. This greater observed siltation rate means that the useful life for the NSD is going to be shorter that expected. Siltation rates for the SSD could not be located, but it can be assumed that they have been underestimated as all the other dams have shown.

Waterlogging

Large-scale irrigation schemes in India have been shown to cause water logging. Rich soils in the states of Punjab and Haryana have been robbed of their use because of water logging. The Indian Institute of Science estimates that 40 percent of the command area for the NSD will become waterlogged. A similar estimate can be expected for the SSD, because both command areas contain, "black cotton soils which are particularly prone to water logging under perennial irrigation". Water logging from the overall Narmada Valley Dam project has already been observed as, "irrigation from the first two dams in the scheme has caused soils to become waterlogged and crop yields to fall".

Destruction of Forests

The destruction of forests is another inevitable consequence of both projects. It can be seen that of the close to 40,000 hectares to be inundated by the SSD, close to 35% of it will be forest land, and of the close to 90,000 hectares to inundated by the NSD, close to 44% of it will be forest land. The MEF has policy in place that requires that for every hectare of forest destroyed, an equivalent hectare of trees must be planted in a non-forest area, or twice as much degraded forest lands can be replanted instead. Since there is not much land even for resettlement let alone forestation, officials have proceeded with the second option of reforestation. Critics, however, cite the fact that, "it is the duty of the Forest Department to afforest degraded forest lands anyway," and that, "the department of Environment and Forests has not even examined whether afforestation can be carried out in degraded forest lands". In recent years, pressure from environmentalist has caused the government of Gujarat to agree to afforest 5,000 hectares in the Kutch region of Gujarat, and to plant 30 million trees along the irrigation canals and on the shores of the reservoirs of the dams. The destruction of forests is a direct consequence of the dams, but of equal importance is the destruction of wildlife in those forests. The scope of the environmental catastrophe that will be caused when entire species of life will be wiped out by these dams is beyond this paper. All that can be said is that officials have not made any plans of adequate relocation for people, let alone wildlife.

Water - Borne Disease

Water- borne disease is a health hazard that is related to large-scale irrigation projects such as the SSD and NSD. India has had experience with water-borne disease as shown by the increased incidence of malaria in the command area of the Tawa project. The Madhya Pradesh Council of Science and Technology has predicted that, "the incidence of malaria, filaria, cholera, gastroenteritis, viral encephalitis, goitre and some other water borne diseases is likely to increase," due to the SSD and NSD projects. Apparently there are measures that could be taken against these water-borne diseases, although no examples could be found. This assumption is made evident by a report to the World Bank that states, "the existing threat from malaria within the irrigation area is serious, and the project has not built in 'appropriate safeguards'". This statement implies that there are "appropriate safeguards" that could be applied and should be if the project is to be of benefit.

Economic Implications of the Project

How much will the Projects Cost and Where has Funding Come From?

Both the SSD and NSD projects have costs that go beyond financial costs of the dams themselves. Some of these costs include the loss of forests and wildlife, the loss of agricultural lands due to waterlogging, the loss of land submerged, the costs associated with resettlement of oustees, and the costs of the irrigation canal system. It is difficult to obtain estimates for such things as the monetary costs of the loss of forests and wildlife, and so the costs that have been estimated will be discussed.

The World Bank has provided funding in the form of loans for the SSD project, and figures for this project are readily available. Funding for the NSD has not come from the World Bank and as a result, the NSD's costs are not readily available, but since the two dams are similar in size the costs for the dam alone can be assumed to be of the same magnitude as the SSD. Estimated costs for the SSD have climbed steadily over the past two decades. In 1983, authorities gave the World Bank an estimate of Rs. 42,020 million ($1,400 million U.S.) for the dam and irrigation canal system, while in 1985 the

World Bank estimated the cost at Rs. 136,400 million. An estimate of Rs. 342,000 million was cited in, "A 1994 World Bank publication". Funding for both projects has been provided primarily by the state governments involved in the projects and the Government of India, with the exception of the SSD, which has had funding provided by both the Indian governments and by a loans from the World Bank and the Japanese government.

The cost of relocation and "human suffering" is a great one that has in fact been estimated. The Narmada Valley Dam Authority (NVDA), which oversees the overall damming of the river, has estimated the "human suffering" by, "determining two years annual income of the family, multiplied by a factor of 1.5," and this has yielded an estimate of Rs. 200 million. This was estimated prior to 1987, and since the estimated costs of the dams have increased since that time, this estimate can be assumed to have increased since then. The Indian government's Department of Forests has also given an estimate of the "social cost" of the projects at about Rs. 90 million.

The unseen environmental costs that will likely arise from these projects have been identified in this paper. One such cost is that of medical care for those inflicted with water-borne diseases from the irrigation canal system. Officials do not consider this when calculating the cost of the projects, and this could prove to be a major cost. The environmental and social costs of the projects have lead to loan contributors pulling out of the SSD project. For instance, the Japanese government cancelled funding for the SSD after concluding that, "even under heroic assumptions, the largest dam being built on the Narmada, the Sardar Sarovar, is environmentally destructive and uneconomic".

What has The World Bank's Involvement been in these Projects?

World Bank funding for the SSD project has amounted to Rs. 8,400 million and the bank has officially pulled out of the project as of 1993. In 1985 the bank committed Rs. 13,500 million, but following increasing pressures from environmental and human rights agencies, the World Bank commissioned a four member review committee in 1991 to investigate the SSD project. Ten months later, the review committee submitted a 363-page report that commented on the resettlement and environmental aspects of the project. The report cited that, "neither the Indian governments nor the World Bank carried out adequate assessments of the human impact of the Sardar Sarovar," and that, "there was no recognition of the special needs of the tribal populations involved." In addition the report stated that, "the compensation packages for MP, Maharashtra, and some Gujarat 'oustees' did not meet WB standards." With regards to the environment, the report stated that, "impact studies were inadequate and that no plan had been prepared to mitigate environmental damage," and that, "the existing schemes for compensatory afforestation, prevention of water logging and salinification in the command area, distribution of drinking water, malaria control came under strong criticism". These concerns brought to light by the review committee lead the World Bank to ask the Indian government to cancel its loan for the project, and it did in so in 1993. With the cancellation of the World Bank loan, critics have said that, "a cash shortage has meant that there is no way the project will be completed before the year 2010, if at all".

Conclusion

The NSD and SSD projects have complex social, environmental and economic implications for the citizens in the dams' target area. The interplay of these implications is very complex and makes it difficult to assess the projects' benefit to development.

Both projects have been designed to provide water for the irrigation of severely drought prone areas of Gujarat and Madhya Pradesh, and the estimate that 40 million people will benefit from this water and the hydroelectric power provided by the dams is a considerable one. The state of Gujarat also endured great financial burden during the droughts of the 1980's, and the dams could potentially save the state millions of dollars.

Who will benefit and suffer from the dams also needs to be considered. Some have said that wealthy farmers and factories depending on cash crops will benefit, while the poor who need relocation will suffer. It is difficult to say if it is only the rich who will benefit because other factors might justify the building of sugar processing factories in the command area, such as cheaper labour. It is clear that a large number of tribals will suffer from the dams because they will have to relocate out of the area to be submerged and this will disrupt their livelihood. Resettlement plans are not adequate and have not even satisfied World Bank regulations, especially for the tribals who have special needs. Since the estimation of the Narmada River's flow has been overestimated, and since reducing the dams' heights will greatly reduce the area to be inundated, dam authorities should seriously consider lowering the proposed heights of both the NSD and the SSD.

The environmental implications of the dams are critical to the evaluation of the dams. Sedimentation and siltation will no doubt reduce the dams' useful life as siltation rates have been underestimated for the NSD and for many other dams in India. The land the surrounds the dams and the irrigation canals is prone to waterlogging and some land will lose its agricultural use, as has been the case in the states of Punjab and Haryana. Forests will be inundated and as a result, a large population of wildlife will be destroyed. Plans for reforestation seem to be inadequate and regulations for reforestation only seem to shift the responsibility of the reforestation of degraded forests from the MEF to the dam authorities, instead of actually creating new forests. Authorities have also neglected to address the problem of water-borne diseases such as malaria. Even the World Bank's independent review strongly criticized the dam authority's plans for disease control.

All of these social and environmental implications add to the cost of the dam projects. The cost of resettlement and "human suffering" is very high and contributes negatively to these projects. The loss of agricultural lands due to waterlogging and the loss of this land in addition to forests due to submergence have not been incorporated into the stated costs of the project. No one has even tried to estimate the associated costs due to the increased incidence of water-borne disease, and these costs are undoubtedly high. The unaccounted costs in the plans for the projects, and the SSD project in particular, have lead to the withdrawal of major capital lenders for the project. As a result of the World Bank's independent review, the bank stopped its funding for the project. This withdrawal of the World Bank, in addition to the withdrawal of the Japanese government, are key indicators for the Indian Central and state governments that the projects have not been adequately assessed and that re-planning needs to take place.

Weighing all of the costs that have been identified against the potential benefit of the projects to development and for irrigation and electricity for over 40 million people is difficult. Some can argue that the benefit of 40 million outweighs the costs for relocation and human suffering to an estimated 250,000. Others can argue that the "human" costs of suffering override any benefits that are potentially available from the projects. An even others (and they already have) can argue that the costs to the environment are too great to justify the benefits. No absolutely correct answer can be offered for the question of the benefit of the project. What can be said is that there are many factor involved and they all are complexly related. It is obvious that mistakes have been made, and people are paying for it right now. The building of the SSD is at a standstill because the Supreme Court of India wants guarantees for the relocation plans of oustees. If assessments had been made *before* the plans for the projects were approved, then modifications may have been made in time, such as the lowering of the dams' heights.

Table 1.1. Villages and Population to be affected by submergence due to the Sardar Sarovar and Narmada Sagar dams

Dam	*State*	*District*	*Villages*	*Population*
Sardar Sarovar	Gujarat	Bharuch Baroda	19	30000
	Maharashtra	Dhulia	33	
	Madhya Pradesh	Dhar	80	
		Jhabua	26	70000
		West Nimar	76	
Narmada Sagar	Madhya Pradesh	Kandwa	167	
		Dewas	39	170000
		Hoshan-gabad	48	
Total			**488**	**270000**

The future of these projects is unclear, but it is clear that they will not be completed on time. Are they beneficial to development? They may be, but at a great cost. These costs should have been considered before construction of the dams, and hopefully the governments within India and the World Bank will learn to complete assessments before any future projects are undertaken.

Table 1.2. Annual Rate of Silting (Hecatre/Metres of Silt) per 100 km^2 of Catchment Area

Project	*Assumed siltation rate while dam was built*	*Observed in surveys*
Malthon	1.62	13.10
Mayurakshi	3.61	16.43
Ramganga	4.29	18.19
Ghod	3.61	15.24
Beas Unit 2	4.29	14.29
Ukai	1.47	10.95
Narmada	1.55	5.62
Tawa	3.61	11.15

Table 1.3. Total Area in Hectares to be affected due to submergence by the Narmada Dams

Project	*Total Submergence*	*Forest*	*Agricultural Land*	*Number of Villages*
Sardar Sarovar	39134	13744	11318	234
Narmada Sagar	91348	40332	44363	254
Total	130482	54076	55681	488

CNG IN DELHI

In 1981, Delhi had just over 500,000 vehicles. In 2000 the vehicle population grew to around 3 million, with two thirds of the vehicles two and three wheelers. Also, more than 10,000 buses ply the roads in Delhi. To curb vehicular pollution, the Supreme Court of Delhi mandated the use of alternative fuels, such as compressed natural gas (CNG) and liquefied petroleum gas (LPG) in all commercial vehicles. In just the past fourteen months, more than 50,000 natural gas vehicles are now operational. LPG vehicles are currently being used in demonstration projects only. The government is also now mandating five percent blend of ethanol in gasoline.

The network of CNG stations is also quite impressive although technical and safety issues abound. As of January 2002, 87 stations exist, including 20 mother stations, 15 daughter booster stations, 36 daughter stations and 16 online stations. By March 2002, 94 stations should be operational. Existing compression capacity is 4.6 lakh kg/day and average sale is 3.7 lakh kg/day. CNG is 40% cheaper than diesel and 68% cheaper than gasoline.

Some stations have terrific signage that showcases to the public the benefits of CNG.

Areas of Need

Unfortunately, some of the technology utilized has been proven unsafe or unreliable through past American experience. Also, lines for the CNG fuel can sometimes wind through the streets for miles. Waits of 3-4 hours to fuel are common. On average a vehicle is only getting a fill of 2000 PSI. This is due to both technical design inadequacies and a calculated decision to reduce operating pressures after a handful of highly-publicized vehicle explosions.

Future supply of natural gas to meet the growing number of vehicles is a concern although gas industry officials express optimism about increased future supply of CNG through successful offshore gas well exploration and expansion of pipeline capacity.

Conversions are an ongoing problem. Kits provide questionable environmental benefit and integration of systems are poorly engineered. As a result, safety concerns about these installations are considerable. For example, brackets were observed to be loosely mounted on steel cylinders, resulting in observable abrasion damage to the cylinder walls.

Training of technicians appears to be lacking. Only Tata and Ashok Leyland, the two Indian original-equipment CNG bus manufacturers, are training mechanics on CNG buses. But training is currently only offered once and is not on-going. While Delhi Transport Corporation (DTC) assured our delegation during a maintenance facility tour that all specified maintenance is performed according to manufacturer specifications, observed conditions of the buses suggest that current maintenance and inspection is inadequate.

Inspection of cylinders is currently not conducted other than on Delhi Transport

Corporation (DTC) Buses, and even here best practices appear lacking. The development of safety standards is a must. Private bus operators are developing their own procedures without regard to safety – such as shutting off cylinders and then reopening them as needed. (Ashok Leyland is now producing CNG buses with roof-mounted tanks on lowfloor buses, although most of the current rolling stock has the cylinders mounted under the chassis, another source of concern.)

Natural gas is only odorized in some area, such as gas lines that are also serving residential customers. In buses, there are 72 joints to check for gas leaks. Obviously, odorized gas would help to detect leaks and cut down on the cost of leak detector equipment, which often isn't being used. However, leak detection sensors are a secondary issue – bus designs must be upgraded. With the use of industry-approved connectors, most of the leak concerns will be mitigated.

Also, government officials remain skeptical of the benefits of CNG and are being fed misinformation by the detractors. For instance, the Chief Minister of Delhi, Sheila Dikshit, said on two occasions during our visit that it is imperative to understand the health impacts of CNG – as if to say that the health benefits of CNG have not already been quantified. Also, detractors are spreading rumors that CNG vehicles emit more ultra fine particles than diesel. She seems to have bought into these arguments and states that she is afraid that CNG will cause serious illnesses. The Centre for Science and Environment (CSE) in New Delhi is really the only voice of reason in this debate. More scientific data must be introduced from other reputable sources to assist CSE in arguing public policy issues. Also, real over the road emission testing of CNG vehicles is needed.

At the same time, many acknowledge that air pollution has improved dramatically in Delhi since the CNG buses were introduced.

The needs identified for CNG also apply to LPG. The only difference is that with LPG the U.S. Department of Energy still has the opportunity to address station design, vehicle integration and standards issues BEFORE a large-scale program is launched. The government has mandated the introduction of LPG as a transportation fuel and ordered domestic oil companies to open 250 LPG fueling sites. Assistance in sanctioning proper LPG kits manufacturers is necessary and the training of technicians must be offered as the program grows.

Steps to Provide Assistance

1. Training: One of the goals of the trip was to launch a train-the-trainer program with the Society of Indian Auto Manufacturers (SIAM). On January 18, 2002, the National Alternative Fuels Training Consortium (NATFC) of the U.S., SIAM and the DTC signed a Memorandum of Understanding "to encourage cooperation among the three institutions to work for setting up a National Alternative Fuels Training Institute in India." The Chief

Minister of Delhi witnessed the signing. DOE's Clean Cities International Program has provided $85k to NATFC to transfer materials and send trainers to Delhi, India.

SIAM will be the lead organization in organizing the following courses:

1. CNG Cylinder Inspection Course – 1 or 2 days in length
2. Train the Trainer on CNG – 5 days in length
3. General Awareness on CNG/LPG for Government Officials – 1 day in length
4. Train the Trainer on LPG – 4 days in length

Based on priorities, SIAM will offer the course to government officials in the Spring 02, followed by the train the trainer and cylinder inspection courses in early Fall 02. DTC will offer classroom and shop space for the training. Technicians working on buses, taxis and auto rickshaws will be targeted for the train the trainer courses on CNG. Inspectors of the Transport Department of Delhi and DTC will be trained on CNG cylinder inspections.

Everyone agrees that training is a critical need. However, based on the many issues identified above, the resources that NATFC bring to the table are only capable of addressing a few of the training and assessment needs - mostly in the technician and operator training areas. When it comes to system assessments, standards development and service station design, other experts are required to address these needs.

2. Delhi Clean Cities: Another goal of the trip was to launch the idea of a Delhi Clean Cities coalition. One of the benefits of a grassroots coalition is that all interested stakeholders come to the table to discuss problems and work on solutions in building a sustainable and safe alternative fuels program. Training is one such program that can be guided by the coalition. Stakeholders identified during the trip: SIAM, DTC, Transport Department of Delhi, Winrock International, Gas Authority of India, Ltd., Association of State Road Transport Undertaking, Maruti, Society of Automotive Engineers – Northern

India Section, Centre for Science and Environment, TATA Engineering, Ashok Leyland, Bharat Petroleum Corporation, Cummins India Ltd., Confederation of Indian Industry, and Indraprastha Gas Ltd.

Stakeholders yet to be identified include private bus operators and taxi associations. Also, appropriate Federal agencies should be encouraged to join.

3. Technical Information Exchange: The third goal was to identify key individuals who should be a part of the Clean Cities International and Gas Technology Institute's Reverse

Technology Tour to be held in October 2002 in conjunction with the International Natural Gas Vehicle Conference. With the help of U.S. Asian Environmental Partnership (US AEP), Clean Cities International is now prepared to organize this tour.

Also, Clean Cities International has asked US AEP to help bring three individuals to Oklahoma City to the National Clean Cities Conference in May 2002.

Meetings and Additional Outcomes of Trip

In addition to the U.S. delegation speaking at the "Seminar & Panel Discussion on Alternate Energy Driven Vehicles" and attending the "International Workshop on High Capacity Bus Systems" at the 6th Annual Indian Auto Expo, the following meetings were conducted:

1. S. Behuria, Director of Marketing and I. Srinivas Rao, Chief Manager of Supply & Logistics, Bharat Petroleum Corporation, Ltd.

   Bharat Petroleum has been mandated by the government to build 42 LPG stations in six Indian cities and 5-6 stations in Delhi. None have been built to date and only company cars have been converted. Opportunities exist for U.S. station equipment manufacturers and conversion kit companies. Training was also discussed. DOE should also talk to Ford/GM about OEM LPG vehicles. (Mr. Dorfman to follow-up with World LPG Association.)

2. Sudam Maitra, Maruti UDYOG Limited; R.B. Madhekar, SAE Secretary—Northern India Section; and A.D. Sindwani, SAE—Northern India Section. The U.S. delegation explained initiatives in India. Mr. Maitra arranged for us to visit a large Maruti manufacturing facility.

3. Mr. Nandi, Mr. Gandi and Sugato Sen, Society of Indian Auto Manufacturers The delegation worked on the details of implementing the MOU. See above. SIAM is responsible for developing a list of questions that can be used to facilitate the development and modification of training modules. (Prof. Frisz to follow-up.)

4. Staff of Transport Department, Government of NCT of Delhi. (Mr. Anand Khullar, Additional Director, was unavailable due to a death in the family.)

   This agency is responsible for inspecting vehicles. Only 10 inspectors are employed to inspect the nearly 3 million light-duty vehicles plying the streets of Delhi. They have not been trained on CNG. Staff members were eager for training.

5. Tarun Das, Director General and V Raghuraman, Senior Advisor—Energy

   Confederation of Indian Industry (CII) SIAM members are also part of CII. CII agreed with the necessity of training in this area. CII will be meeting with Secretary of Energy in February and will possibly meet with Ms. Rood. CII has a lobbyist in Washington and is a powerful stakeholder.

6. Ted MacDonald, EPA is funding an inspection and maintenance pilot program and helping with emission

modeling. It is possible to tie the training programs together. EPA supports the MOU.

7. Shyamala Abeyratne, President and Saroj Mishra, Sr. Program Officer, Winrock.

   International India

   Winrock is promoting large-scale ethanol production and use in India. The government is mandating five percent ethanol blends, in which they need to produce 400 ml to meet demand. Ethanol is currently made from molasses. Winrock has gathered key state officials together at several workshops. They would make an excellent stakeholder in the Delhi Clean Cities coalition and have numerous contacts in the various states, which can be useful as the CNG/LPG programs grow outside of Delhi.

8. A.S. Lakra, Director, Association of State Road Transport Undertakings

   This association is comprised of 67 Transport Corporations around India. They provide professional development and training to their members. They also have 500 bus vendors as members and they develop certification of parts and monitor performance. They operate the Central Institute of Road Transport where they train managers and engineers. It is also a testing facility. They will have an annual conference in late March-April and would like materials on CNG and a possible speaker.

9. Rajeev Khanna, General Manager and Santosh Kumar, General Manager (Training), Gas Authority of India (GAIL) and A.K. De, Managing Director, Indraprashta Gas Limited A.K. De is responsible for the CNG stations in operation. They had many technical questions, such as CNG station design and technology upgradation, natural gas quality, CNG filling system, development of safety standards, and are in need of training and information. A list of needs was given to the delegation.

10. Visited a DTC Bus Depot – services Ashok Leyland buses.

    Classroom and service areas are available for training. More than 50 people attended and listened to Prof. Frisz in earnest. Seven bus fires have occurred and leaks are of utmost concern.

11. Anumita Roychowdhury, Centre for Science and Environment
    Centre for Science and Environment is the lead NGO that pushed for a CNG vehicle mandate in Delhi. The delegation exchanged information and gained support for the initiatives. Concern was expressed that Delhi already has a lot of institutes so a coordinated effort is important.
12. R.K. Jaiswal, Senior Manager of TATA Engineering and Debabrata
    Chakraborty, Cummins Power Systems India Ltd.
    More than 1600 technicians have been trained and 1200 TATA drivers have been trained. However, Mr. Jaiswal expressed concern for establishing a training program that is on-going. TATA is willing to provide the trainers for the train-the-trainer course. They have trained drivers how to refuel but drivers are not assigned to buses.
    14 stations have been authorized to provide parts for the CNG buses. Currently, there is no facility for maintenance on privately owned buses.
    Cummins will begin to sell lean-burning CNG engines in India. Three parts will be shipped from the U.S. to India to be assembled into the CNG engine.
    Rob Adams, a consultant for DOE's domestic Clean Cities Program, was hired by TATA to examine quality control issues and safety of TATA's buses.
    (Mr. Dorfman has already followed up with Mr. Adams and now has a better understanding of the consulting work that he performed on behalf of TATA.)
13. Maruti Plant Tour – The plant is capable of producing 500,000 vehicles annually.
    They produce a CNG van and will be developing a LPG sedan. They are interested in quality control. (Dorfman to follow-up.)
14. Ron Sissem of Louis Berger Group, Greenhouse Gas Pollution Prevention
    Project Mr. Sissem is spearheading a transportation planning project in Hyderbad, India.

This past fall three delegates visited the Transtar manufacturing facility in Denver, where they manufacture CNG/hybrid buses. Officials in Hyderbad would like to demonstrate LPG/hybrid technology. (Ms. Rood to follow-up with Denver coalition)

BHOPAL GAS TRAGEDY

Introduction Around 1 a.m. on Monday, the 3rd of December, 1984, in a densely populated region in the city of Bhopal, Central India, a poisonous vapor burst from the tall stacks of the Union Carbide pesticide plant. This vapor was a highly toxic cloud of methyl isocyanate. Of the 800,000 people living in Bhopal at the time, 2,000 died immediately, and as many as 300,000 were injured. In addition, about 7,000 animals were injured, of which about one thousand were killed. "A series of studies made five years later showed that many of the survivors were still suffering from one or several of the following ailments: partial or complete blindness, gastrointestinal disorders, impaired immune systems, post traumatic stress disorders, and menstrual problems in women. A rise in spontaneous abortions, stillbirths, and offspring with genetic defects was also noted." (The Bhopal Disaster) This incident we now refer to as the Bhopal Gas Tragedy, which has also been called "Hiroshima of the Chemical Industry" one of the worst commercial industrial disasters in history.(Cohen)

After the incidence, over the next few years, numerous studies were conducted, many theories were explored, and the involved parties accused each other. In this chapter, I will try to explore the various causes offered for the tragedy. In the course of my research for this case study, I came across many articles that put blame on various people and groups involved in the tragedy. I found one document particularly interesting from a rhetorical standpoint. This document, titled Union Carbide: Disaster at Bhopal , was authored by the retired Vice President of Health, Safety and Environmental Programs in Union Carbide Corporation. So for this paper, I would also like to rhetorically analyse this document and also, try to explore the various image restoration strategies that Union Carbide Corporation used through the course of the crisis.

The Tragedy: Possible Causes The post-accident analysis of the process showed that the accident started when a tank containing methyl isocyanate (MIC) leaked. MIC is an extremely reactive chemical and is used in production of the insecticide carbaryl. It is presumed that the scientific reason for the accident at Bhopal is that water entered the tank where about 40 cubic meters of MIC was stored. When water and MIC mixed, an exothermic chemical reaction started, producing a lot of heat. As a result, the safety valve of the tank burst because of the increase in pressure. This burst was so violent that the coating of concrete around the tank also broke. It is presumed that between 20 and 30 tonnes of MIC were released during the hour that the leak took place. The gas leaked from a 30 m high chimney and this height was not enough to reduce the effects of the discharge. The reason was that the high moisture content (aerosol) in the discharge when evaporating, gave rise to a heavy gas which rapidly sank to the ground. The weather egged on this process. The conditions on the fateful day were typical for a clear night in the region, with a weak wind which frequently changed direction, which in turn helped the gas to cover more area in a shorter period of time (about one hour). The weak wind and the weak vertical turbulence caused a slow dilution of gas and thus allowed the poisonous gas to spread over considerable distances. (Chemical Accidents...)

Many different terms have been used to describe the events in Bhopal that early morning of December 3, 1984: accident, disaster, catastrophe, crisis and also as sabotage, conspiracy, massacre, and experiment, whichever best suited the arguments that would help to pin the 'blame' on somebody. In his book titled The Bhopal Tragedy: Language, Logic and Politics in the Production of a Hazard, the author William Bogard "Each of these descriptions, in its own way, minimizes the problem of human agency and intention, and thus refuses to address directly the issue of responsibility." Bogard goes on to point out that the best way to describe this incidence would be a tragedy because, "In calling Bhopal a tragedy, we are still permitted to say that intention and agency were involved in how the event unfolded and that responsibility must ultimately rest with someone or some group. But unlike saying that Bhopal was the deliberate result of sabotage, a conspiracy, or some diabolical experiment involving human guinea pigs- charges that are virtually impossible to prove in any case- a tragedy, in contrast, emerges out of a complex of confused and misguided intentions, many of which may be honorable in themselves but when forged to the actual chain of events produce the worst possible outcome." In the last twenty years, numerous studies have been conducted on the incident and there are numerous deductions. In most of the studies, the two main agencies analysed were the Union Carbide Corporation and the Indian Government of the Late Prime minister Rajiv Gandhi and the Madhya Pradesh state government of Arjun Singh. One of the main reasons for the tragedy was found to be a result of a combination of human factors and an incorrectly designed safety system. "A portion of the safety equipment at the plant had been non-operational for four months and the rest failed. When the plant finally sounded an alarm—an hour after the toxic cloud had escaped—much of the harm had already been done."(The Bhopal Disaster). Union Carbide itself believed the theory that the tragedy resulted when " a disgruntled plant employee, apparently bent on spoiling a batch of methyl isocyanate, added water to a storage tank"(Browning). Still others, like the many experts in industrial safety, believe that the tragedy was preventable, arguing that it was the due to "....the negligence on the part of the Union Carbide Corporation and its corporate subsidiary Union Carbide of India Ltd.(UCIL), which had the responsibility for taking care of the day-to-day operations of the facility". The corporation and its subsidiary were also charged with corporate irresponsibility for pursuing the profits instead of the safety and hazard standards. The Madhya Pradesh State government had not mandated any safety standards and Union Carbide failed to implement its own safety rules, apparently comfortable in the knowledge that it was not contravening Indian regulation. "The Bhopal plant experienced six accidents between 1981 and 1984, at least three of which involved MIC or phosgene, a highly poisonous gas used in World War I and a component in the manufacture of MIC. The accidents were generally small scale- one worker was killed in 1981- but official inquiries required by law were often shelved or tended to minimize the government's or the company's role". It is noted that it was probably this pattern of neglect that failed to bring about the much needed change in the malfunctioning safety equipment and improperly trained workers at the chemical plant. Even so, this negligent behavior on the part of Union Carbide regarding safety standards raised little concern among the citizens of Bhopal. So, why were the people of Bhopal so indifferent when voicing their concerns on the safety factors in the Union Carbide plant? Why was nothing done about the defective safety equipment? To understand this, it is important to understand that India is a poor nation. The country needed pesticides to protect her agricultural production. MIC is used to produce pesticides that control insects which would in turn, help increase production of food - central to India's Green Revolution, which was ironically, US imposed. Initially, India imported the MIC from the United States. In an attempt to achieve industrial self-sufficiency, India invited Union Carbide to set up a plant in the state of Madhya Pradesh to produce methyl isocyanate. The license was given to them on the belief that the chemical industry would provide the desperately needed jobs and capital for the people of the country. To the people of the city of Bhopal, Union Carbide was a highly respected , technically advanced Western company that would bring them the jobs they needed. This coupled with political power and scientific expertise worked together to changed the people's perception of what was dangerous and more importantly what was safe.

The Analysis: Union Carbide's Reaction Through all the months immediately following the incident, Union Carbide never directly apologized to the Indian government and her people or to the people of Bhopal. The Indian Government, in response to the tragedy and pressure from the Indian people, filed a compensation lawsuit against the UCC for an estimated $3 billion. On the other hand, Union Carbide strongly felt that the Indian government was to blame. This was the headlines in The New York Times on Dec. 17, 1986, The Union Carbide Corporation in Dec., 1986, while continuing to deny liability, filed a counter suit against the government of India and the State of Madhya Pradesh regarding the 1984 disaster at Carbide's Bhopal subsidiary. The company is charging the governments with "contributory" responsibility for the leak of poisonous gases, saying both governments knew of the toxicity of methyl isocyanate but failed to take adequate precautions to prevent a disaster. The government of India has sued Union Carbide for at least $3 billion in compensation for the victims of the leak of methyl isocyanate.(D4) This was iterated in the document titled Union Carbide: Disaster at Bhopal , written by Jackson B. Browning for the Union Carbide Corporation. At the time of the Bhopal tragedy, Jackson B. Browning was the Vice President responsible for the Health, Safety, and Environmental Programs in the corporation. He was one of the spokesmen for the corporation during the crisis in 1984 and was also in charge of the teams that responded to and investigated the tragedy. He retired from UCC in 1986.

Browning's document outlines the various aspects of the Bhopal tragedy from the perspective of the Union Carbide Corporation. In the very second paragraph on page one of the article, the author notes that the cause for the accident, as believed by the parent company- "Although it was not known at the time, the gas was formed when a disgruntled plant employee, apparently bent on spoiling a batch of methyl isocyanate, added water to a storage tank."(Browning). This was the main argument by UCC in their defense and they still maintain the same. The corporation needed to divert the blame for the tragedy from themselves to something or somebody else, especially one that would catch the attention of anybody remotely interested in the incident. They used "sabotage". What I found interesting was that on the one hand Browning called the incident of December 3rd a "massive industrial disaster" and on the other hand, a premeditated action- a sabotage. To me, the two don't fit together. Disaster would mean 'even if we knew, we could have done nothing about it' and sabotage on the other hand would mean 'if the process had not been tampered with, there would have been no leak, no loss of life'. But this was clearly not the case. Studies conducted on this incidence by a Dr. Paul Shrivastava, tell a completely different story. Dr. Paul Shrivastava, an Associate Professor of Business in New York University and Executive Director Industrial Crisis Institute Inc., NY conducted studies that revealed that Bhopal was neither an isolated incident nor the first of its kind in the corporation. There had been many accidents of similar nature in UCC's American plants prior to the Bhopal accident. He found that 28 major MIC leaks had occurred in UCC's West Virginia plant during the five years preceding the Bhopal incident, the last one occurring only a month before. His studies found that the 'sabotage' theory was UCC's way to avoid paying the huge amount that the Indian government had demanded as settlement. Interestingly, UCC, till date, has been unsuccessful in presenting any evidence to prove that theory and has never disclosed the name of the supposedly guilty employee.

All the previous accidents in the other Union Carbide plants were not highly publicized events, and hence, there were no repercussions that UCC had to face. But in the case of the Bhopal tragedy, the magnitude of the incident worked against them and made it difficult for them to distance themselves from it. In his article, discussing the theories of image restoration, Benoit notes that there are two components to an attack on one's image: an undesirable act has been committed and you are responsible for that action. "Only if both of these conditions are believed to be true by the relevant audiences is the actor's reputation at risk...". The Bhopal incident was too big for the public to ignore and added to everything else, there was a huge loss of human life. This naturally drew attention. Even Browning notes in his article that " the scope of the Bhopal tragedy made it to "page one" material in the weeks and months that followed." Union Carbide was under attack from all sides as news of the leak spread and they needed to make arguments to achieve one particular goal- "restoring or protecting their reputation". Benoit argues that ".when our reputation is threatened, we feel compelled to offer explanations, defenses, justifications, rationalizations, apologies or excuses for our behavior". But these defenses and excuses needed to be made to the audience that mattered the most to you. To the Union Carbide Corporation in the United States, the audiences were the people around them in the US and the media.

The Press seemed to be the main focus in Browning document. The tone of the document suggested that the main audiences to pacify would be the media and once that was done the corporation would have definitely 'saved face'. The other relevant audiences needed to be identified carefully. In the document, Browning has a sub-section titled "Keeping Vital Audiences Informed". Under this section, Browning himself clearly identifies that audiences they were responding to: the most visible-the media, and other interested parties like the customers, shareholders, suppliers and other employees. Nowhere in this whole section was there a mention of the people of India or the people of Bhopal. There seemed an urgency for the corporation to assure the people of the United States who were their main stockholders, that such an incident would not happen here. Browning notes in the document that Warren Anderson, the then chairman of the UCC, and he were summoned to appear before the House Commerce and Energy Committee to answer one question- "Can it happen here?" It seems like this pretty much proved that the process of image restoration for the corporation was not all that difficult because of the large distance between the 'vital' audiences and the site of the disaster Bhopal. In his discussion of the tragedy, Benoit notes, The unusual aspect of Carbide's public image is the fact that the public believed Union Carbide was responsible for Bhopal and had not told the truth about it- yet had a generally favorable overall opinion about the company. This may reflect a partial lack of interest in events that occurred in distant lands and suggests that salience of the accusations to the audience of an important factor of image restoration. This shows an important factor of restoring one's image in the eyes of the public depends to a great extent on how relevant the unfavorable event is in their eyes, in other words, how close to home is the tragedy.

As the first step towards image restoration, Browning's main strategy seems to have been to distance the corporation from the site of the disaster. Browning, very early in the document, points out that the Union Carbide Corporation had only 50.9% stake in the affiliate, the Union Carbide India Ltd. He also makes clear that all the employees in the company were Indians and that "...the last American employee at the site had left two years before." Union Carbide Corporation maintained that it did not have any hold over its Indian affiliate. The UCC argued that the day-to-day working in UCIL was independent of the parent company and hence it was not to be held responsible. But most of the research showed that this was not really true. In spite of denials, it appears the Union Carbide company in Danbury, Connecticut had substantial authority over its affiliate. Many of the day to day details, such as staffing and maintenance, were left to Indian officials, but every major decision, such as the annual budget, had to be cleared with the American headquarters, and directives were often issued from the US. And in addition to this, by both Indian and US laws, a parent company (UCC in this case) holds full responsibility for any plants they operate through subsidiaries and in which they have the majority stake. Hence, it seemed like the main aim of making an argument that UCIL was independent could be for two purposes: 1) to avoid paying the large sum of $3,000 million that India demanded as compensation or 2) to find a 'scapegoat' to divert the blame onto. In his article, Keith Michael Hearit, an Assistant Professor in Communication Studies, Purdue University, discusses the concept of scapegoating with respect to saving face, " instances in which corporations cannot deny the validity of the charges, they are forced to deal with the issue of guilt and responsibility to restore their social legitimacy. At such time, corporate apologist offer individual/group dissociations. An individual/group dissociation is a scapegoating strategy by which a rhetor seeks too transfer guilt to another." In this case, UCC, by noting that UCIL had an all Indian workforce and the last American employed had left two years before, attempted to restore its image by differentiating the affiliate from the rest of the organization. This is one of the many modes of image restoration discussed by Benoit.

Union Carbide used some of these strategies to restore its reputation after the gas leak in Bhopal. One of the strategies employed is that of 'corrective action' and stands out in an interesting section of Browning's document titled "Safety Emphasized". Under this section, Browning tries to established that Bhopal was a stray incident and should not be held against the corporation because "No analysis of Union Carbide's reaction to the Bhopal tragedy is possible without recognizing the considerable emphasis the company and its affiliates had placed on safe operations"(Browning). To lend credibility to the corporation's cause, Browning cites an international management specialist, Dr. Richard Robinson, a professor in Massachusetts Institute of Technology, commenting on the tragedy as saying that Union Carbide was one of those multinational corporations who were very dedicated to the safety aspect of their plants and that " it is particularly depressing that it was Union Carbide which was involved". Browning notes that considering the company's strict safety policies that the news of the Bhopal tragedy was "astounding". Arguments in this section are more devoted to explaining that it would be unfair to assume that the accusations that UCC was not careful with safety, were true. It is, in a sense, a form of apologia- the corporation is utilizing the "act/essence dissociation. An act/essence dissociation distances the apologist from the wrongdoing by arguing that while the wrongdoing admittedly occurred, it was an isolated act that does not represent the apologist's true nature." And this is often followed by the next step -to point out how the corporation has worked to remedying the unfortunate incidence. This is exactly what Union Carbide Corporation did.

Under the two sections titled "First Steps At Control" and "Contingency Planning and Experience Help", Browning lists out all the things that UCC did immediately following the first call they got about the tragedy. He notes that vital decisions were made - the UCC facility making MIC in the US was shut down; a task force led by the chairman of UCC, Warren Anderson, was set up; medical and technical teams were dispatched to the site of the tragedy "within 24 hours". He also noted that "Union Carbide had a contingency plan for emergencies" The people of UCC worked together, with the press and the 'vital audiences', in to help in dealing with the "terrible facts of the tragedy". What is interesting is that most of the research done on the incident points to the fact that Union Carbide did not have any kind of emergency plans in its Indian subsidiary. So much so that when the accident occurred and people started pouring into the hospitals in Bhopal complaining about the various ailments, the hospital staff had on idea of what had happened or what to do. "The city health officials had not been informed of the toxicity of the chemicals used at the Union Carbide factory. There were no emergency plans or procedures in place and no knowledge of how to deal with the poisonous cloud."

Browning ends his document noting, with confidence, that the approach used by UCC at the time of the disaster were in his opinion "correct ones". He also notes that today's Union Carbide Corporation is a very different company. The Corporation now works twice as hard on its safety operations and that "money and staff were committed to those objectives".

Conclusion Thirteen years later not much has changed. Union Carbide India Ltd. is an abandoned site in Bhopal. UCC sold its share of the affiliate. In October of 1991, the Indian Supreme Court upheld a settlement, which had been appealed from a lower court decision of 1989, under which Union Carbide had to pay $470 million in compensation of all claims. In 1996, at Union Carbide's annual meeting, William H. Joyce, its chief executive, declared that the company had no intention of doing anything further for the victims. This resolve was apparently reversed, as the company announced that it is planning to support the building of a $ 20 million hospital for the victims of the Bhopal tragedy through a London based independent charitable trust. The construction should be complete in mid-1998 and will be operational by the end of this year.

Today, Union Carbide is a six billion dollar company, whose worldwide sales percentage is increasing every financial year. It seems like their image restoration strategy worked for them. "Union Carbide may have been aided in this matter by an unconscious ethnocentric bias in the public. It is reasonable to assume that if this terrible tragedy had occurred here in the United states (rather than in a foreign country), its image would have suffered even more."

Bhopal was one of the worst industrial disasters in history. For all its horrors, the tragedy had at least one beneficial consequence- the intense public debate that followed the tragedy made more private citizens aware of the hazards of the chemical industry as a whole. It put the lethal nature of the chemical industry in out in the open. In response to this, the Chemical Manufacturing Association created the 'Responsible Care Program' that is now being implemented worldwide in at least 22 countries. The Program's aim is to improve community awareness, emergency response and employee health and safety.

Bibliography

Alvares, C. and R. Billorey. 1987. Damning the Narmada: The Politics Behind the Destruction. The Ecologist 17(2): 62-73.

Beegel Susan F., Susan Shillinglaw, and William N. Tiffney Jr., eds. *Steinbeck and the Environment*. Tuscaloosa: University of Alabama Press, 1997.

Berry Thomas. *The Dream of Earth*. San Francisco: Sierra Club Books, 1988.

Blake Nelson M. *Land into Water – Water into Land: A History of Water Management in Florida*. Tallahassee: University Presses of Florida, 1980.

Cahn Robert. *Footprints on the Planet: A Search for an Environmental Ethic*. New York: Universe Books, 1978.

Coyle David Cushman. *Conservation: An American Story of Conflict and Accomplishment*. New Brunswick, NJ: Rutgers University Press, 1957.

Crosby Alfred W., Jr. *The Columbian Exchange: Biological and Cultural Consequences of 1492*. Westport, CT: Greenwood Press, 1972.

Daily Gretchen C., ed. *Nature's Services: Society's Dependence on Natural Ecosystems*. Washington, D.C.: Island Press, 1997.

Damned by its own words. 1992. The Economist 324(Sept. 5): 36.

de J. E. Steiguer *Age of Environmentalism*. New York: McGraw-Hill, 1997.

Devall Bill, and George Sessions. *Deep Ecology: Living as if Nature Mattered*. Salt Lake City: Gibbs Smith, 1985.

Dharmadhikary, S. 1996. Sardar Sarovar Project Update. Online: International Rivers Network, http://www.irn.org/irn/programs/narmada/update960929.html.

Dixon, J. A., L. M. Talbot, and G. Le Moigne. 1989. Dams and the Environment. Washington: The World Bank.

Esterbrook Gregg. *A Moment on Earth: The Coming of Age of Environmental Optimism*. Penguin, 1995.

Esteva, G., and M. S. Prakash. 1992. Grassroots Resistance to Sustainable Development: Lessons from the Banks of the Narmada. The Ecologist 22(2): 45-51.

Fleagle Robert G. *Global Environmental Change: Interactions of Science, Policy, and Politics in the United States*. Westport, CT: Praeger, 1994.

Frederiksen, H. 1996. Case Study: Narmada Basin Development in India. Journal of Water Resources Planning and Management 122(March/April): 86-87.

Glickson Artur. *The Ecological Basis of Planning*. Ed. by Lewis Mumford. The Hague: Nijhoff, 1971.

Hattangadi, S., and T. Ichniowski. 1992. Megaproject in India under attack. ENR 228:10-11.

Hays Samuel P. *Beauty, Health, and Permanence: Environmental Politics in the United States, 1958-1985*. New York: Cambridge University Press, 1987.

Helvarg David. *The War Against the Greens: The "Wise-Use" Movement, the New Right and Anti-environmental Violence*. San Francisco: Sierra Club Books, 1994.

Indian Supreme Court Halts Construction on Sardar Sarovar. 1996. Online: International Rivers Network, http://www.irn.org/irn/pubs/wrr/9601/narmada.html.

Jayaraman, K.S. 1988. Narmada Valley irrigation plan a test case for World Bank. Nature 335: 583.

Krech William III. *The Ecological Indian: Myth and History*. New York: W.W. Norton, 1999.

List Peter C. *Radical Environmentalism: Philosophy and Tactics*. Belmont, CA: Wadsworth, 1993.

Lodaya, K. 1994. Exercising the Other Option. Online: International Rivers Network, http://www.irn.org/irn/programs/narmada/alt2.html.

McCully, P. 1994. Sardar Sarovar Project (SSP) An Overview. Online: International Rivers Network, http://www.irn.org/irn/programs/narmada/irnoverview940525.html.

McCully, P. 1997. Earthquake Hits Narmada Valley. Online: International Rivers Network, http://www.irn.org/irn/pubs/wrr/9706/narmada.html.

Merchant Carolyn. *Major Problems in American Environmental History*. Lexington, MA: Heath, 1993.

Muir John. *Our National Parks*. New York: AMS Press, 1970. Reprint of 1901 ed.

Olmsted Frederick Law Jr., and Theodora Kimball, eds. *Frederick Law Olmsted: Landscape Architect, 1822-1903*. Bronx, NY: Benjamin Blom, 1970. Reprint of *Forty Years of Landscape Architecture: Being the Professional Papers of Frederick Law Olmsted, Senior* (1928).

Petulla Joseph M. *American Environmental History*. 2nd ed. Columbus, OH: Merrill, 1988.

President's Council on Sustainable Development. *Sustainable America: A New Consensus for Prosperity, Opportunity, and a Healthy Environment*. Washington, D.C.: President's Council on Sustainable Development, 1996.

Rubin Charles T. *The Green Crusade: Rethinking the Roots of Environmentalism*. New York: Free Press, 1994.

Sale Kirkpatrick. *The Green Revolution: The American Environmental Movement, 1962-1992*. New York: Hill and Wang, 1993.

Sattaur, O. 1991. Fair deal denied to people displaced by dam. New Scientist 131(Aug.5): 10.

Sattaur, O. 1991. Greens in muddy water over Indian dam. New Scientist 132(Oct.5): 16.

Sen, A. 1995. We shall not be moved. New Scientist 147(Aug.5): 51.

Sessions George, ed. *Deep Ecology for the Twenty-first Century: Readings on the Philosophy of the New Environmentalism*. Boston: Shambala, 1995.

Shabecoff Philip. *A Fierce Green Fire: The American Environmental Movement*. New York: Hill and Wang, 1993.

The World Bank. 1995. India: Recent Economic Developments and Prospects. Washington: The World Bank.

Thorne-Miller Boyce, and John Catena. *The Living Ocean: Understanding and Protecting Marine Biodiversity*. Washington, D.C.: Island Press, 1991.

Thorne-Miller Boyce, and John Catena. *The Living Ocean: Understanding and Protecting Marine Biodiversity*. Washington, D.C.: Island Press, 1991.

Tokar Brian. *The Green Alternative: Creating an Ecological Future*. San Pedro, CA: Miles, 1987.

Ward Barbara, and Rene Dubos. *Only One Earth: The Care and Maintenance of a Small Planet*. New York: W. W. Norton, 1972.

Wilson Edward O. *Naturalist*. New York: Warner Books, 1994.

Wood, J. 1993. India's Narmada River Dams: Sardar Sarovar Under Siege. Asian Survey 33: 968-984.

World Commission on Environment and Development. *Our Common Future*. New York: Oxford University Press, 1987.

Worster Donald. *Nature's Economy: A History of Ecological Ideas*. New York: Cambridge University Press, 1977.

Young, J.E. 1994. Drowning Out Protest: India's Colossal Mistake. World Watch 7(July / August): 2.

Index